IONIZATION METHODS IN
ORGANIC MASS SPECTROMETRY

RSC Analytical Spectroscopy Monographs

Series Editor: Neil W. Barnett, *Deakin University, Victoria, Australia*

The series aims to provide a tutorial approach to the use of spectrometric and spectroscopic measurement techniques in analytical science, providing guidance and advice to individuals on a day-to-day basis during the course of their work with the emphasis on important practical aspects of the subject.

How to obtain future titles on publication

A standing order plan is available for this series. A standing order will bring delivery of each new volume immediately upon publication. For further information, please write to:

Turpin Distribution Services Ltd., Blackhorse Road, Letchworth, Hertfordshire SG6 1HN, UK. Telephone: +44(0) 1462 672555; Fax: + 44(0) 1462 480947

RSC
ANALYTICAL
SPECTROSCOPY
MONOGRAPHS

Ionization Methods in Organic Mass Spectrometry

Alison E. Ashcroft

Formerly of Micromass UK Ltd., Tudor Road, Altrincham, Cheshire, UK

Now with the Centre for Biomolecular Sciences, School of Biochemistry & Molecular Biology, University of Leeds, Leeds, UK

THE ROYAL
SOCIETY OF
CHEMISTRY
Information
Services

A catalogue record for this book is available from the British Library

ISBN 0-85404-570-8

© The Royal Society of Chemistry 1997

Published by The Royal Society of Chemistry,
Thomas Graham House, Science Park, Milton Road, Cambridge CB4 4WF, UK

Typeset by Computape (Pickering) Ltd, Pickering, North Yorkshire, UK
Printed by Bookcraft (Bath) Ltd

Preface

The aim of this monograph was to produce an introductory guide to ionization methods which could be referred to on a daily basis during the practice of organic mass spectrometry. There are numerous ionization methods available to the modern organic mass spectroscopist, and it can be difficult to choose the most appropriate one for the analysis in question. This book attempts to describe the main features of these methods so that the mass spectroscopist can decide which to use for a particular application, and much of the information provided herein has been transposed into readily accessible tabular form to meet this aim.

Although the book was not intended to be a treatise on mass spectrometers or mass spectrometry in general, a brief introduction was deemed necessary if only to clarify nomenclature and highlight which instruments can be used with the various ionization techniques. I make no apology for omitting references to Ion Trap mass spectrometers which are also used very successfully with many of the ionization methods described; my reasoning is that as I have had no practical experience of this type of mass spectrometer, I am not qualified to advise others how to use them!

After the introductory chapter, the remaining chapters are each dedicated to a particular ionization method, some more popular than others in modern times. For each method of ionization, there is a list of common application areas, a short description of the technique, and a section on how to set up and obtain the best performance with the method in question. Finally some examples of sample analyses are highlighted. The references for each chapter are certainly not intended to be a complete literature search in that particular area; they are simply supplied as examples of different aspects of the ionization methods (my favourites if you like). The reason for this is twofold; not only would a literature search covering thousands of references be quite out of place in a book of this size, it would almost certainly be out of date before the book was printed. Most mass spectroscopists have access to good library facilities and it is recommended that a literature search is performed at the time that it is required to generate the most up-to-date references.

As a practising mass spectroscopist for 14 years, I have tried to create the type of book that I would have welcomed over the years; not too bulky a treatise, enough theory to enable one to understand a method so that it can be used successfully, but not so much that may unnerve a relative newcomer to mass spectrometry. After all, mass spectrometry, at least in the author's opinion, is a practical analytical technique, and the whole point in having a mass spectrometer is to use it, and to use it well. Hopefully this book will help users get over the initial hurdle of dealing with sometimes complicated equipment and become sufficiently proficient to solve real, analytical problems.

Acknowledgements

I would like to thank my employers, Micromass UK Ltd., for allowing me to use data for many of the figures, and in particular Dr Charles Smith for reading through the manuscript. I would also like to thank colleagues past and present from both Micromass UK Ltd. and Kratos Analytical Ltd. for providing me with much beneficial advice over the years. Lastly I would like to thank Bill and Helen for their support during this work.

Alison E. Ashcroft
January 1997

Contents

CHAPTER 1

Introduction

1 An Introduction to Mass Spectrometers

Although it is beyond the scope of this book to delve deeply into the theory and physics of mass spectrometers, a brief introduction would appear to be necessary, not only to clarify the nomenclature used for the various techniques and hardware described in the remainder of this monograph, but also to encourage the reader to turn to more complete texts on the subject.

A mass spectrometer, like Caesar's Gaul, can be divided into three fundamental parts, namely the **ionization source**, the **analyser**, and the **detector** (see Figure 1.1). Mass spectrometers are used primarily to provide information concerning the **molecular weight** of a compound, and in order to achieve this, the sample under investigation has to be introduced into the ionization source of the instrument. In the source, the sample molecules are ionized (because ions are easier to manipulate than neutral species) and these ions are extracted into the analyser region of the mass spectrometer where they are separated according to their **mass (m) to charge (z) ratios (m/z)**. The separated ions are detected and the signal fed to a data system where the results can be studied, processed, and printed out. The whole of the mass spectrometer (except for Atmospheric Pressure Ionization sources) is maintained under vacuum to give the ions a good chance of travelling from one end of the instrument to the other without any interference or hindrance. Nowadays the entire operation of the mass spectrometer and often the sample introduction process are usually under complete data system control and the operator hardly needs to move away from the computer terminal to perform the sample analyses.

Many **ionization methods** are available and each has its own advantages and disadvantages. The method of ionization used depends on the sample under investigation, the type of mass spectrometer being used, and the available equipment. This book describes the more commonly encountered ionization methods, and aims to provide an account of their set-up and basic operation. Once the ionization method has been set up and has been shown to be operating at its optimum performance, then the operator can start to develop the technique for the particular samples under scrutiny. The optimum performance of any ionization method will depend on the performance and condition of the mass spectrometer, the reliability of any other equipment and materials involved, including gas and liquid chromatographs and chromatography columns, the purity of any solvents or gases used, and the quality of the

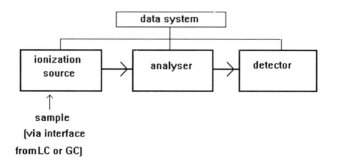

Figure 1.1 *Simplified schematic diagram of a mass spectrometer*

standard samples. However, it should be remembered that the optimum performance *must* be established *before* any sample analyses are undertaken, and this performance should be verified every day, or more frequently if laboratory procedures dictate or if problems are suspected. If the performance is not as good as expected then steps should be taken to retrieve any losses in sensitivity or resolution.

As well as there being a good choice of ionization methods, there are also many different ways of introducing samples into the ionization source depending on the ionization method being used and the type of samples under investigation. For example, single-substance samples can be inserted directly into the ionization source by means of a probe whereas complex mixtures will benefit from some kind of chromatographic separation *en route* to the ionization source, and this could involve interfacing liquid chromatography (LC), gas chromatography (GC), supercritical fluid chromatography (SFC), or capillary electrophoresis (CE) to the mass spectrometer. The methods of interfacing to the various ionization methods are described in more detail in the relevant chapters.

After the ionization source, the ions proceed to the **analyser** region, and a mass spectrometer is generally classified by the type of **analyser** it accommodates. There is a variety of **analysers**, and the ones referred to in this book are those that are most frequently encountered in organic mass spectrometry, namely the **magnetic sector**, the **quadrupole**, and the **time-of-flight**. Each will be discussed in a little more detail later in this Chapter (see Chapter 1, Section 2). Not all ionization methods are compatible with all of these analysers, as will be revealed where appropriate in the text.

The **detector** could be one of several possibilities including *inter alia* photomultipliers, electron multipliers, microchannel plates, and diode array detectors. On a day-to-day basis, the detector gain should be set at the appropriate level for acquiring data.

The remainder of this Chapter aims to provide a brief overview of a range of mass spectrometers and indicate which ionization methods are appropriate. I have tried to summarize the various ionization methods, instruments, and sample introduction methods in several different ways so that these summaries

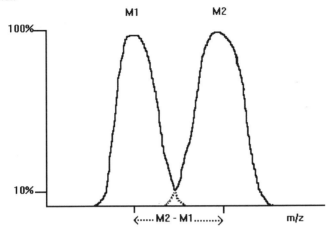

Figure 1.2 *Mass resolution illustrated with the 10% valley definition*

can be referred to at a later date when reading the more detailed chapters to help put the various topics in perspective. The summaries may appear to overlap, and if this is so, I apologize; it was simply my intention to display the data in a readily accessible manner, emphasizing different significant aspects so that the text would appeal to a variety of readers.

2 The Mass Analyser

Introduction

Resolution

The main function of the **mass analyser** is to separate, or **resolve**, the ions formed in the ionization source by their mass to charge ratios (***m/z***).

The **resolution** (***R***)[1] of a mass analyser, or its ability to separate two peaks, is defined as the ratio of the mass of a peak (***M_1***) to the difference in mass between this peak and the adjacent peak of higher mass (***M_2***) (see Figure 1.2), *i.e.*:

$$R = \frac{M_1}{M_2 - M_1} \qquad (1.1)$$

where R = resolution,

M_1 = the mass of a peak, and

M_2 = the mass of an adjacent, higher mass peak.

In the simplest terms, a singly charged ion at *m/z* 1000 could be separated

---

[1] W. H. McFadden, *Techniques of Combined Gas Chromatography/Mass Spectrometry: Applications in Organic Analysis*, Wiley-Interscience, New York, 1973.

This book is recommended for its detailed explanation of resolution, and also its descriptions of different mass analysers.

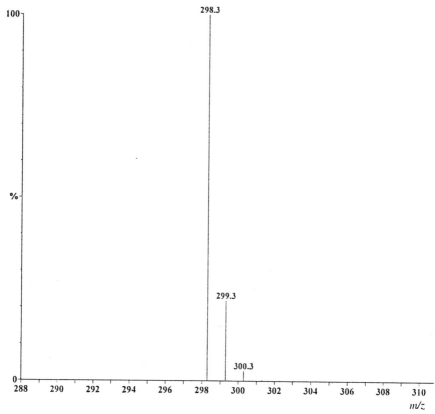

Figure 1.3 *Molecular ion ($M^{+\bullet}$) for the compound of molecular formula $C_{19}H_{38}O_2$, showing the isotope distribution*
(Reproduced with permission from Micromass UK Ltd.)

from another singly charged ion at *m/z* 1001 if a resolution of 1000 is available. Similarly, a singly charged ion at *m/z* 2000 would require a resolution of 2000 to separate it from a second singly charged ion at *m/z* 2001, whereas a singly charged ion at *m/z* 100 would need only 100 resolution to separate it from another singly charged ion at *m/z* 101.

Resolution, when referring to magnetic sector mass spectrometers, is often described by the '**valley definition**' where a 'resolution of 10% valley' (see Figure 1.2) means that two peaks of equal intensity are considered resolved when the height of the valley between the peaks is 10% of the peak height. Alternatively, and less frequently, one may allude to a resolution of 50% valley. Quadrupole and time-of-flight mass spectrometers are generally less able to provide high (or better than unit) resolution, although recent advances with time-of-flight instruments have led to improvements. In such cases, a peak width can be described instead; for example, one might say the sample was analysed with a peak width of 0.5 amu measured at half of the maximum height of the peak, or 0.5 amu **FWHM (full width half maximum).**

Isotope Distributions

In general the resolution actually required for most analyses is such that the singly charged **isotope patterns** of the detected ions are readily discernible, and for applications involving molecular weights *ca.* 1500 da or less, this can be provided by magnetic sector, quadrupole, and time-of-flight mass spectrometers.

If one considers a small organic compound of molecular formula $C_{19}H_{38}O_2$, then under electron impact (EI) ionization conditions (see Chapter 3) with unit resolution set for the analyser, a **molecular ion** ($M^{+\bullet}$) is generated at *m/z* 298 (see Figure 1.3) which relates to the intact molecules (less one electron) in which all the atoms are the lowest mass (and in this case the most abundant) isotopes (*i.e.* ^{12}C, ^1H, and ^{16}O). This value can be taken to be the molecular weight of the compound. There will also be lower intensity ions at *m/z* 299, which correspond to molecules of the same compound in which one ^{12}C atom has been replaced by a less abundant, and therefore less probable, ^{13}C isotope. The relative intensities of these two ions should relate to the natural abundances of the isotopes multiplied by the number of carbon atoms in the molecule. In other words, the intensity of the *m/z* 299 ion compared to the *m/z* 298 ion should be equal to 1.11 (because the natural abundance of ^{13}C is 1.11% of the natural abundance of ^{12}C) multiplied by 19, which equals 21.09%. For higher molecular weight samples which contain more carbon atoms, the probability of one of the ^{12}C atoms having been replaced by a ^{13}C atom increases, and indeed when the number of carbon atoms in a molecule reaches 90, it becomes more probable to find a molecule with one ^{12}C atom replaced by a ^{13}C atom, than to find a molecule with all its carbon atoms of the ^{12}C type. The isotope distribution for a compound of theoretical molecular formula $C_{100}H_{202}$ is shown in Figure 1.4 to illustrate this.[2]

If the sample under investigation is already known, then the theoretical molecular weight can be calculated from the molecular formula of the compound. If the average atomic masses from the periodic table are used for this purpose, an accurate, but **average molecular weight** of 298.5095 daltons (da) results for the above sample of molecular formula $C_{19}H_{38}O_2$. If unit resolution has been set, this will *not* be the mass of the ion detected and reported by the mass spectrometer. Remember that because mass spectrometers separate ions according to their *m/z* ratio, so the isotopes of the atoms should be taken into account when calculating the molecular formula of a compound. The *dominant* ion in this particular molecular weight cluster is the $^{12}C_{19}{}^1H_{38}{}^{16}O_2$ ion, whose accurate but **monoisotopic molecular weight** is 298.2872 da. Figure 1.5 presents a list of some of the most commonly encountered atoms together with their monoisotopic and average masses.

If a mass spectrometer has been properly calibrated, then the mass accuracy

[2] D. H. Williams and I. Fleming, *Spectroscopic Methods in Organic Chemistry*, McGraw-Hill Book Company (UK) Ltd., Berkshire, UK, 2nd edn, 1973.
This book provides a good basis not only for an explanation and examples of isotope patterns, but also for general spectral interpretation.

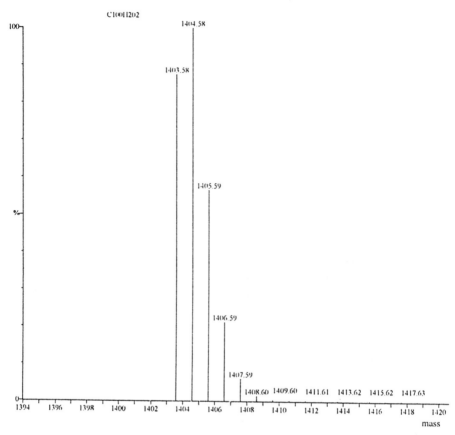

Figure 1.4 *Theoretical isotope distribution for the molecular formula $C_{100}H_{202}$* (Reproduced with permission from Micromass UK Ltd.)

should be good to at least 0.1 da. In the sample described previously, $C_{19}H_{38}O_2$, the difference between the average and monoisotopic molecular weights is not great and indeed both have the same nominal mass; in this case the spectrum could have been interpreted equally well regardless of whether the operator had used monoisotopic or average values for the calculations. This is not always the case though, and so care should be exercised, especially when dealing with high molecular weight samples (> 2000 da), or with samples that exhibit irregular isotope patterns such as those containing chlorine, bromine, or transition metal atoms such as nickel and zinc. As an example, if the average and monoisotopic accurate masses are calculated for a sample of molecular formula $C_{18}H_{12}Cl_2FNO_4S$, values of 428.2674 and 426.9848 respectively are obtained. Now there is a significant difference between the two calculations, and a mass spectrum that produced a molecular ion at *m/z* 427 would quite correctly be consistent with the monoisotopic calculation, but would indicate (mistakenly) that the sample was not as expected if the average masses had been used.

Atom	Isotope	Natural abundance	Monoisotopic mass	Average mass
hydrogen	^{1}H	99.985	1.0078	1.0079
	^{2}H	0.015	2.0141	
carbon	^{12}C	98.90	12.0000	12.0110
	^{13}C	1.10	13.0034	
nitrogen	^{14}N	99.63	14.0031	14.0067
	^{15}N	0.37	15.0001	
oxygen	^{16}O	99.76	15.9949	15.9994
	^{17}O	0.04	16.9991	
	^{18}O	0.20	17.9992	
fluorine	^{19}F	100	18.9984	18.9984
sodium	^{23}Na	100	22.9898	22.9898
phosphorus	^{31}P	100	30.9738	30.9738
sulfur	^{32}S	95.02	31.9721	32.0660
	^{33}S	0.75	32.9715	
	^{34}S	4.21	33.9679	
	^{36}S	0.02	35.9671	
chlorine	^{35}Cl	75.77	34.9689	35.4527
	^{37}Cl	24.23	36.9659	
potassium	^{39}K	93.26	38.9637	39.0983
	^{40}K	0.01	39.9640	
	^{41}K	6.73	40.9618	
bromine	^{79}Br	50.69	78.9183	79.9040
	^{81}Br	49.31	80.9163	
iodine	^{127}I	100	126.9045	126.9045
caesium	^{133}Cs	100	132.9054	132.9054

Figure 1.5 *Some frequently encountered atoms with their monoisotopic and average atomic masses*[3]

The two halides chlorine and bromine each have two isotopes separated by two mass units; chlorine consists of ^{35}Cl and ^{37}Cl in the approximate ratio 3:1, and bromine consists of ^{79}Br and ^{81}Br in approximately equal ratios. This produces in both cases a distinctive and readily recognisable pattern which is a good aid for compound identification. If a compound has more than one bromine or chlorine atom, or one or more of each, then the isotope pattern increases in complexity and distinction, as shown in Figure 1.6.

Finally, the expected isotope pattern for an organometallic compound of molecular formula $C_{24}H_{54}Br_2NiP_2$ is illustrated in Figure 1.7 to give an idea of the complexity involved with some samples, and to emphasize the necessity for correctly calculating the masses of the isotopes in order to be able to interpret the data properly.

[3] J. R. De Laeter, K. G. Heumann, R. C. Barber, I. L. Barnes, J. Cesario, T. L. Chang and T. B. Coplen, *Pure Appl. Chem.*, 1991, **63**, 975, and references cited therein.

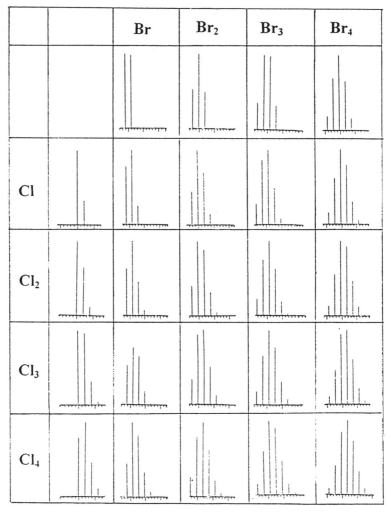

Figure 1.6 *Theoretical isotope distributions for compounds containing multiple chlorine and/or bromine atoms*

Most mass spectrometers will resolve ions with unit resolution up to at least 2000 da, and so monoisotopic atomic masses are used in these cases. Above 2000 da, the resolution should be checked and if it is insufficient to resolve adjacent isotopes, then average atomic masses are used in calculations.

Accurate Mass Measurements

Occasionally the nominal molecular weight of a sample, as determined with an accuracy of say, 0.1 da from the mass spectrum, is not sufficient to characterize the sample. This is especially true if the sample is an original one whose

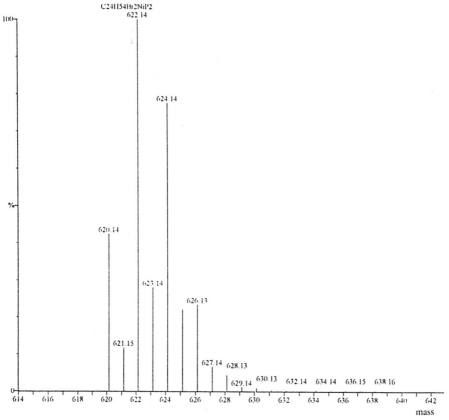

Figure 1.7 *Theoretical isotope distribution for the organometallic compound of molecular formula* $C_{24}H_{54}Br_2NiP_2$
(Reproduced with permission from Micromass UK Ltd.)

molecular formula has to be validated, or if there is a chance that the sample could have one or more structures which have the same nominal but different accurate masses. For example, the formulae $C_{21}H_{36}O_3$ and $C_{19}H_{32}N_2O_3$ have monoisotopic masses of 336.2664 and 336.2413, respectively. The mass spectrum for this sample could indicate a molecular weight for the compound of 336.2 da, but from this information, it is not possible to say which formula is the correct one; both fit the data equally well. Therefore an **accurate mass measurement** is required, which should provide a measurement within 5 parts per million (ppm) error of the correct answer. Accurate mass measurements require due care and attention in their operation. A suitable reference material needs to be used, and a means of maintaining the reference in the source simultaneously with the sample must be sought. For electron impact analyses a volatile reference material such as heptacosa is often admitted into the ionization source through a permanently sited reference inlet, whilst the sample is introduced into the source by means of a probe, or eluting from a GC column.

If two alternative formulae can be proposed from the nominal molecular weight obtained from the mass spectrum of the compound, and if both are expected to be present in the same sample, then the resolution required for their *m/z* separation should be calculated so that the resolution on the mass spectrometer can be set before the experiment is initiated. In the example above, the resolution *R* needed is given by:

$$\frac{336.2413}{336.2664 - 336.2413} = 13\,396$$

The mass spectrometer should be set to provide at least this amount of resolution if the experiment is to separate these two structures. In general, resolution above 2000 (10% valley definition) requires the use of a magnetic sector mass spectrometer.

Accurate mass measurements can be made at any resolution; resolution is the criterion to be considered when separating masses.

Methods of Using the Mass Analyser

There are different methods of acquiring data when using a mass spectrometer, and these should be taken into account when designing an experiment. The most usual method of acquiring data is by scanning the mass analyser over an appropriate *m/z* range, thus producing a **mass spectrum** from which (hopefully) molecular or quasimolecular (molecular related) ions will provide an indication of the molecular weight of the sample. If the sample has fragmented (fallen to pieces) in the ionization source, then these ions will also have been collected, and often the fragment ions can be studied and information regarding the structure of the sample pieced together. Almost all samples are analysed with a **full scanning** experiment initially to produce as much information as possible about the sample, and full scanning acquisitions are possible with magnetic sector, quadrupole, and time-of-flight mass spectrometers. Under appropriate conditions, accurate mass measurements can also be carried out.

If the analyst is investigating known compounds which have been character-ized previously, and wants to ascertain whether or not the expected compound is present, or needs to determine the concentration level of the sample in a biological or ecological matrix, then often a **selected ion recording (SIR)** analysis is performed. Before this can be carried out, one or more significant and characteristic ions from the sample must be specified in the acquisition parameters. These ions could be the molecular or quasimolecular ions, for example, and/or intense, diagnostic fragment ions. The mass analyser will then monitor the specified ions by switching from one to the next. This technique is more sensitive than a full scanning one, because all the available time is spent on the ions of interest rather than monitoring all the ions over a stipulated *m/z* range. The sensitivity is highest if only one ion is monitored, but care must be taken to ensure that no other isomeric or isobaric compounds are present in the same sample. A good compromise is to monitor two or three ions for each compound under scrutiny, as this gives good sensitivity while providing more credence to the results.

SIR acquisitions are often performed in the pharmaceutical industry where low levels of drugs and metabolites need to be ascertained in complex biological matrices which give rise to a high level of background ions. Both magnetic sector and quadrupole mass spectrometers are used for SIR analyses but not, in general, time-of-flight instruments. Magnetic sector mass spectrometers, with their high resolution capabilities, can also perform SIR at high resolution whereby the accurate, monoisotopic mass ions are specified and monitored, thus producing very much more specific results. **High resolution SIR** is used in the field of dioxin analysis, for example.

By far the best method of performing SIR is to use a means of sample introduction, such as liquid or gas chromatography, which generates sample peaks of relatively short peak widths that can be integrated, as opposed to the probe methods of sample introduction which deliver the sample into the ionization source at a near constant rate over long periods of time.

Magnetic Sector Mass Spectrometers

If a mass spectrometer is considered as comprising a source, an analyser, and a detector, then the mass spectrometers described in this particular section all have a magnet as the analyser. **Magnetic sector mass spectrometers** can have simply a magnet, or (more frequently) a magnet together with an **electrostatic analyser (ESA)**, and in the latter case the magnet can either be followed by or preceded by the electrostatic analyser.

The magnet serves to separate the ions produced in the ionization source and in this case the separation is achieved by magnetic deflection. In order to pass the ions from the ionization source into the magnetic analyser, the source is held at a high voltage, typically between 2000 V and 8000 V, which causes acceleration of the sample ions out of the source with a high velocity. The effect of the magnetic field is to deflect the ions in a curved trajectory. The ions of smaller mass are deflected more than those of larger mass. For an ion to reach the detector at the end of the mass spectrometer, it must follow a path of a certain radius (r) through the magnetic field (of strength B), Figure 1.8. The equation for the path of the ions through the magnet is as follows:[4]

$$m/z = \frac{B^2 r^2}{2V} \qquad (1.2)$$

where m = mass of an ion,
$\quad z$ = the number of charges on the ion,
$\quad B$ = the strength of the magnetic field,
$\quad r$ = the radius of curvature of the ion's path, and
$\quad V$ = the accelerating (source) voltage.

---

[4] J. R. Chapman, *Practical Organic Mass Spectrometry*, John Wiley & Sons, Chichester, UK, 2nd edn, 1994.
This book presents full details of the geometry of magnetic sector mass spectrometers.

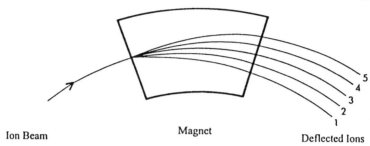

Figure 1.8 *Ion's path through a magnetic analyser of a mass spectrometer*
(Reproduced with permission from Micromass UK Ltd.)

From this equation, it can be seen that if the magnetic field is scanned while the accelerating voltage and the radius of curvature are held constant, then in turn all the ions of different masses will pass through the magnet in succession and emerge from the exit, pass through the collector slit and reach the detector. One scan of the magnet results in the production of one m/z spectrum.

If it is necessary to differentiate between ions that have the same nominal but different exact masses, higher resolution is required, and for this reason most commercial magnetic sector mass spectrometers are usually designed with an electrostatic analyser that operates in conjunction with the magnetic sector to improve resolution (see Figure 1.9). The mass spectrometer is now termed a **double focusing** instrument, and resolutions in excess of 150 000 (10% valley definition) can be achieved on some such instruments.

When the ions exit the ionization source, they will have a spread of energies which contributes to their peak widths. The ESA focuses the velocity, and hence kinetic energy of the ions. The ESA does not mass analyse. The path of an ion through the ESA is expressed by the following equation:[4]

$$\frac{mv^2}{r'} = eE \tag{1.3}$$

where m = the mass of an ion,
  v = the velocity of an ion,
  e = the charge on an electron,
  E = the ESA field strength, and
  r' = the radius of the ion's path in the ESA.

The combination of a magnetic and an electrostatic analyser is termed double focusing because it is both directional (or angular) and energy focusing. A well-designed double focusing mass spectrometer has both high resolution and high sensitivity. Such high specifications often result in an expensive instrument, but for some specific applications, *e.g.* dioxin analyses and high resolution accurate mass measurements, these instruments are irreplaceable and invaluable. The mass range of the magnetic sector instrument depends on

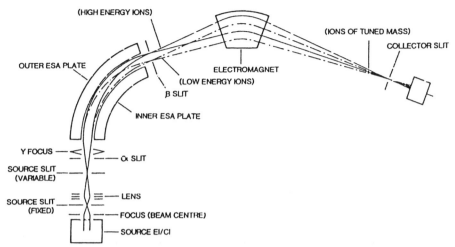

Figure 1.9 *Double focusing magnetic sector mass spectrometer with the magnet preceded by the electrostatic analyser*
(Reproduced with permission from Micromass UK Ltd.)

the design of the magnet and this will vary from one mass spectrometer to another. Although proteins of molecular weight above 20 000 da have been analysed successfully,[5] in general very little is cited in the literature for samples above 10 000 da, and with the advent of electrospray ionization[6] (see Chapter 2), large mass ranges are not now an important issue. Magnetic sector mass spectrometers are often considered to be more difficult to operate then quadrupole and time-of-flight mass spectrometers, and certainly the high voltage source is less forgiving to erroneous usage and more demanding to LC interfacing technology.

Quadrupole Mass Spectrometers

Mass spectrometers with **quadrupole analysers** have the reputation of being easier to use than magnetic sector mass spectrometers, and are popular instruments for a diverse range of applications. Quadrupole mass spectrometers are ideal for coupling with both liquid and gas chromatography and so their usage includes drug metabolism studies, pharmacokinetic analyses, pesticide work, the detection of flavours and fragrances, and many other application areas. Their reliability and robustness makes them the instrument of choice for multi-user systems such as those of the 'open access'[7,8] type.

[5] B. N. Green and R. S. Bordoli, in 'The Molecular Weight Determination of Large Peptides by Magnetic Sector Mass Spectrometry', *Mass Spectrometry of Peptides*, ed. D. M. Desiderio, CRC Press, Florida, USA, 1991.

[6] J. Fenn, *J. Phys. Chem.*, 1984, **88**, 4451.

[7] D. V. Bowen, F. S. Pullen and D. S. Richards, *Rapid Commun. Mass Spectrom.*, 1994, **8**, 632.

[8] L. C. E. Taylor, R. L. Johnson and R. Raso, *J. Am. Soc. Mass Spectrom.*, 1995, **6**, 387.

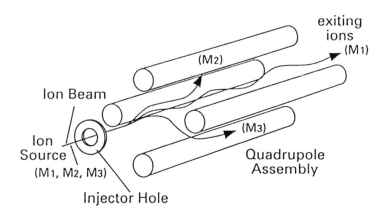

Figure 1.10 *Arrangement of a quadrupole analyser*
(Reproduced with permission from Micromass UK Ltd.)

As with magnetic sector mass spectrometers, electrospray ionization (see Chapter 2), which generates multiply charged ions that are detected at m/z values much lower than the molecular weight of the large biomolecules, has chiefly removed the mass range limitations which traditionally restrict the analysis of singly charged compounds. Quadrupole instruments are limited in their resolution to analyses performed at low (*i.e.* unity) resolution. Their strong points are high sensitivity, ease of use, reliability, and ability to cope with large volumes of solvent (*i.e.* LC coupling) flowing into the ionization source (which is not held at a high potential) for extended periods of time.

The quadrupole analyser, as the name suggests, consists of four parallel, cylindrical poles or rods which are arranged symmetrically, as shown in Figure 1.10.[9] The voltage connections to the rods are such that opposite rods have the same polarity while adjacent rods have opposite polarity. The voltage applied has two components: a **direct current (DC)** component (U) and a **radio-frequency (Rf)** component [$V_0(\cos \omega t)$, where ω = the frequency of the Rf voltage]. The ions produced in the ionization source exit with only a small accelerating voltage (and hence a relatively low energy) and pass down through the quadrupole assembly. On entering the electric field the ions oscillate and, at a certain radio-frequency, ions of a certain mass will be in a state of stable oscillation which enables them to proceed straight through the quadrupole assembly and reach the detector. Under these conditions, all other masses will not be undergoing stable oscillation and will be lost on the rods of the quadrupole, and in this way mass separation is achieved.

In order to produce a mass spectrum, U and V_0 must be varied, whilst the ratio U/V_0 is kept essentially constant. The mass of the ions being analysed at any one time is proportional to V_0, and so a linear increase in V_0 produces a

[9] W. Paul, H. P. Reinhard and U. von Zahn, *Z. Physik*, 1958, **152**, 143.

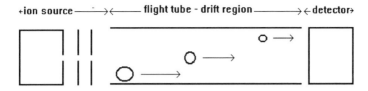

Figure 1.11 *A simplified time-of-flight mass spectrometer*

linear increase in mass. This linear mass scale is relatively simple to calibrate by analysing known standards, and the calibration tends to hold for long periods of time, if the same set of operating parameters is maintained.

The DC voltage is varied to effect a change in the resolution, and the higher the resolution, the fewer the number of ions detected and hence the lower the sensitivity. Resolution *versus* sensitivity is a common compromise in mass spectrometry.

If the DC voltage is switched off, the quadrupole is described as operating in Rf mode only, and all ions will perform stable oscillations if the Rf voltage is sufficiently low. In cases such as this, the quadrupole is not being used as a mass analyser, but as a high transmission lens, and as such has often been employed as a '**collision cell**' situated between two different mass analysers in **tandem mass spectrometers (MS–MS)**. These instruments are described briefly later in this Section.

Time-of-Flight Mass Spectrometers

The **time-of-flight (TOF)**[1] mass analyser has had a revival in recent years after having shown signs of dormancy for a decade or more. The commercial resurrection came about due to the interest aroused by **matrix assisted laser desorption ionization**[10] (**MALDI**; see Chapter 7) in the field of biochemistry, and MALDI is ideally suited to TOF analysis.

In a TOF analyser, a 'pulse' or 'bundle' of ions is accelerated out of the source by a reasonably high voltage, and the ions are then allowed to drift in a field-free region to the detector, see Figure 1.11. The ions all have the same energy and so the lighter ions travel faster than the heavier ions. Thus the separation of the ions depends on the length of time each one takes to travel along the flight tube. A time signal is initiated when the ions are pulsed out of the ion source and also when they arrive at the detector so that the time of flight can be measured. Assuming that the velocity of the ion is determined primarily by the accelerating voltage, then Equation (1.4) can be used to describe the kinetic energy of the ion:

[10] M. Karas and F. Hillenkamp, *Anal. Chem.*, 1988, **60**, 2299.

$$\text{kinetic energy} = \frac{mv^2}{2} = eV \qquad (1.4)$$

where m = the mass of an ion,
$\quad v$ = the velocity of the ion,
$\quad e$ = the charge on an electron, and
$\quad V$ = the accelerating voltage.

The flight time (t) for the ion can be calculated from Equation (1.5),

$$t = \frac{l}{v} \qquad (1.5)$$

where t = the flight time of the ion,
$\quad v$ = the velocity of the ion, and
$\quad l$ = the length of the flight.

By eliminating v from these two equations, the mass of an ion can be calculated if the time of flight, the length of the flight, the charge on an electron, and the accelerating voltage are known, see Equation (1.6). To perform this calculation, the time is measured and the other parameters are all of fixed values,

$$\frac{m}{e} = \frac{2Vt^2}{l^2} \qquad (1.6)$$

where m = the mass of an ion,
$\quad e$ = the charge on an electron,
$\quad V$ = the accelerating voltage,
$\quad t$ = the flight time of the ion, and
$\quad l$ = the length of the flight.

The mass range of a TOF analyser is virtually limitless and so for this reason, the technique has been applied to the analysis of high mass polymeric materials of both biomolecular and synthetic origin. Usually MALDI–TOF spectra exhibit a protonated or deprotonated singly charged molecular ion, depending on whether positive or negative ionization, respectively, was used. Sometimes these ions are accompanied by related ions such as a doubly charged species, or a dimeric species. The resolution of a TOF mass spectrometer depends on the time spread of the ion beam, and unit resolution up to *ca.* 2000 da was, until recently, the standard. Recent advances in technology employ reflectron lenses and 'delayed extraction'[11,12] to improve resolution by minimizing small differences in ion energies, and in these cases up to 12 000 resolution (FWHM) is available.

[11] W. C. Wiley and I. McLaren, *Rev. Sci. Inst.* 1955, **26**, 1150.
[12] J. J. Lennon and R. S. Brown, *Anal. Chem.*, 1995, **67**, 1988.

A reflectron is an ion optic, mirror-like device in which the direction of the flight of the ions is reversed. Ions with greater kinetic energies penetrate further into the reflectron than do those ions with smaller kinetic energies. The ions that penetrate further also spend a longer time in the reflectron, and hence take longer to reach the detector. The overall effect is that the reflectron decreases the spread in the flight times for ions of the same m/z ratio but different kinetic energies, and so improves the resolution. The differences between linear and reflectron modes of operation are described in Chapter 7.

With delayed extraction, the ions in the source are initially allowed to disperse in a near zero electric field, and then at an appropriate time some 0.5 to 10 μs later, an electric field is applied to the ions. As the faster ions, which have travelled longer distances, receive less energy from the applied field than slower ions, velocity focussing is achieved and ions of the same mass arrive simultaneously at the detector regardless of their initial velocities.

TOF mass spectrometers, similarly to quadrupole mass spectrometers, are acclaimed to be straightforward to use and high in reliability and robustness. At the moment they are rarely used in conjunction with on-line chromatography, although they do show an aptitude for the direct analysis of mixtures, as suppression effects from one compound to another appear to be quite low generally.

Tandem Mass Spectrometry

The operation of two or more connected mass analysers in sequence (or tandem) to perform one analysis is known as **tandem mass spectrometry (MS–MS)**. The use of tandem mass spectrometry is to provide further information of a more specific nature about a sample by generating and mass analysing fragment ions from the sample-related ions created in the ionization source. Therefore tandem mass spectrometry is used for the structural elucidation of unknown samples, or for the detection of a known compound in a difficult matrix where specificity is of the utmost importance. The mass analysers can be the same, or mixed, and the principal, commercial varieties are as follows:

quadrupole–quadrupole (tandem quadrupole mass spectrometer);
magnetic sector (including an ESA)–quadrupole (hybrid mass spectrometer);
magnetic sector (including an ESA)–magnetic sector (including an ESA) (four sector mass spectrometer).

Up-and-coming instruments such as a quadrupole–TOF,[13] in addition to TOF mass spectrometers operating with post-source decay[14,15] facilities, should also be considered for structural elucidation MS studies. All of the

[13] A. N. Verentchikov, W. Ens and K. G. Standing, *Anal. Chem.*, 1994, **66**, 126.
[14] R. Kaufmann, B. Spengler and F. Lützenkirchen, *Rapid Commun. Mass Spectrom.*, 1993, 7, 902.
[15] B. Spengler, in 'New Instrument Approaches to Collision Induced Dissociation using a TOF Instrument', *Protein and Peptide Analysis by Mass Spectrometry*, ed. J. R. Chapman, Humana Press Inc., NJ, 1996.

tandem mass spectrometers have a **collision cell**[4] in between the two analysers and the collision cell, in the cases of the tandem quadrupole and hybrid instruments, is often an Rf-only quadrupole (described earlier in this Section), or sometimes a hexapole or an octapole arrangement of rods which are operated in the same way. An inert gas such as argon, helium or xenon is usually admitted into the collision cell where, with sufficient energy, it will bombard any sample ions that have been mass selected by the first analyser, and cause fragmentations to occur.

Several different types of MS–MS experiment can be performed on all of these tandem mass spectrometers. By and large, the results from a quadrupole–quadrupole tandem mass spectrometer are comparable to those obtained on a magnetic sector–quadrupole instrument as both give rise to what are termed **low energy collisions** (*i.e.* the ions enter the collision cell with a low energy), while additional fragmentation can take place in the **high energy collisions** (*i.e.* the ions enter the collision cell with a high energy) which occur when using four-sector mass spectrometers.

Probably the most frequently performed MS–MS experiment is the **'product ion'** scan (or daughter ion scan) and this involves setting up the first mass analyser (be it a quadrupole or a magnet–ESA) to transmit only ions of a certain mass. For example, if the scanning spectrum of a sample has been acquired, maybe a molecular weight of 500 da is indicated from the data. This is useful information and may provide the analyst with sufficient knowledge to ascertain whether or not a reaction or extraction has worked successfully. On the other hand, if the sample was a complete unknown, then it would be useful to have some structural information also. The molecular ions at *m/z* 500 (if the spectrum was acquired under electron impact conditions) are selected and transmitted through the first analyser into the collision cell. Because none of the other ions generated in the ionization source is transmitted, this is very much a specific analysis. In the collision cell, these *m/z* 500 ions collide with molecules of the inert gas and undergo fragmentation. Some chemical bonds are more susceptible to cleavage than others, for example highly polarized C–O and C–N bonds often fragment more readily than (but not to the exclusion of) C–C bonds in the same molecule. All the fragment ions produced enter the second mass analyser, which is scanned over an appropriate *m/z* range so that all of them will be detected. The product ion spectrum generated will probably still contain some of the precursor (or parent) ions, usually with a number of fragment ions, all of which arise *directly* from the fragmentation of the precursor ions.

Alternatively the reverse procedure may be undertaken, and a **'precursor'** (or parent) ion MS–MS scan performed. In this case, the second analyser would be set up to transmit ions of a certain, user-specified mass, while the first analyser would be scanned to allow all ions through into the collision cell. The ions would fragment in the collision cell, but only those that generate the specified fragment would be transmitted by the second analyser and detected. For example, many aromatic compounds give rise to fragment ions at *m/z* 91, corresponding to the tropylium ion $C_7H_7^+$. If the analyst had a mixture containing several aromatic compounds amongst many others, the second analyser could be set to transmit only the m/z 91 ions while the first analyser was scanned, and the spectrum produced should show some evidence for several ions of higher than m/z 91 values, all of which undergo fragmentation to produce directly m/z 91 ions.

A third type of MS–MS experiment involves **constant neutral loss** scans. The loss of a neutral species from a charged ion by fragmentation is quite common; for example carboxylic acids are prone to the loss of a neutral molecule of carbon dioxide (CO_2) which results in the formation of ions some 44 da lower than the precursor ions. For MS–MS constant neutral loss analyses, both the first and the second mass analysers are scanned simultaneously but with a specified offset (equivalent to the neutral species under investigation, in this case 44 da) in mass. In such cases, all the ions generated in the ionization source are transmitted through the first analyser into the collision cell where fragmentation is induced. However only the fragments which differ from their precursors by the specified mass difference are analysed by the second mass analyser and subsequently detected.

The final MS–MS experiment to be mentioned in this Chapter is termed **multiple reaction monitoring (MRM)** or **selected decomposition monitoring**. This type of analysis does not provide any new information about a compound, rather it verifies whether or not a known compound is present. The spectral properties of this compound will have been well established and an experiment devised to make best use of these characteristics.

Suppose a sample has a molecular weight of 430 da, and its electron impact mass spectrum indicates confirmatory molecular ions at m/z 430. Let us presume also that the MS–MS product ion spectrum is dominated by intense fragment ions at m/z 315, together with several other, less intense fragment ions. If the molecular ions are selected uniquely to pass through the first analyser into the collision cell, and the resulting, specific fragment ions at m/z 315 are the sole ions selected for transmission through the second analyser, then this would constitute an MRM experiment. It is apparent that this type of experiment is particularly specific, as very few other compounds would have ions at m/z 430 which fragmented directly to product ions at m/z 315, the exception being some (but not all) related isomers.

MRM presents an alternative, rather more specific experiment to SIR, and should be used if a mass spectrometer with these capabilities is available. MRM detection limits are usually excellent, owing to the high signal-to-noise ratio.

Maintenance

Although it is beyond the scope of this treatise to go into great detail on the subject of maintenance due to the space available and the number of different types of instruments, it must be stressed that mass spectrometers should be regularly maintained, and that the manufacturer's instructions should be followed diligently.

At the very least, a daily check-up should be made on the cleanliness of the ionization source by devising a quickly executed sensitivity test that can be as simple as analysing a known sample and checking the absolute intensity of the ions in the mass spectrum, or on an oscilloscope display (either a traditional model or, more commonly these days, through the data system). If the specification is not attained, or if the performance of the mass spectrometer drops off during sample acquisitions, then it is probable that the source has become dirty and should be cleaned thoroughly. This will mean dismantling the source according to the instructions supplied, cleaning the various parts by appropriate methods, and then reassembling and reinserting the source into the mass spectrometer. Care should be taken to use the correct tools, solvents, emery paper, and alignment devices as the individual source parts are usually of a high precision and can be easily, if unintentionally, damaged.

Daily check-ups should also be made on the state of the vacuum of the mass spectrometer, and the vacuum gauges for both the source and the analyser should be checked so that any deterioration in readings can be rectified. The high vacuum pumps are backed by lower vacuum rotary pumps which should have their oil levels checked weekly and any instructions provided concerning gas ballasting, especially in the case of LC–MS operation, should be strictly adhered to.

It is advisable to keep an up-to-date log book for each mass spectrometer and make a note of any readings, specifications, maintenance, modifications, and repairs which may be carried out. This would also include software revisions, gas cylinder changes, source and analyser tuning parameters, and sample analyses. It is surprisingly difficult to remember all of these items in chronological order at a later date, and a log book such as this becomes even more valuable if the instrument has multiple users.

The last tip I shall mention under this heading is: if in any doubt at all about the correct procedure for any mode of operation or maintenance, then do not hesitate to consult the manual, a more experienced operator, or the manufacturer – or all three!

3 Ionization Methods in Organic Mass Spectrometry

Which Ionization Methods are Compatible with the Mass Spectrometers?

When the mass spectrometer has been selected, it is necessary to choose an appropriate ionization technique. The purpose of this Section is not to go into detail about each ionization method, as the remainder of the book is dedicated to that purpose, but to summarize the individual techniques and put them into perspective with reference to other ionization methods, to the different types of mass analyser, and of great importance, to the samples requiring analysis. It is necessary not only to understand the points of excellence and the shortcomings

Table 1.1 *Ionization methods and their compatibility with different types of mass spectrometers*

Ionization method	Principal ions detected $(+/-)^a$	Mass spectrometer[b]	Sample classes (approx. MW limit)
electron impact EI	$M^{+\bullet}$ and some fragment ions	M, Q	non-polar and some polar organic compounds, $\leq$ *ca.* 1000 da
chemical ionization CI	MH^+ $M^-, (M-H)^-, M^{-\bullet}$	M, Q	non-polar and some polar organic compounds, $\leq$ *ca.* 1000 da
electrospray ES	$MH^+, (M+nH)^{n+}$ $(M-H)^-, (M-nH)^{n-}$	M, Q, TOF	polar organics, proteins, biopolymers, organometallics, $\leq$ *ca.* 200 000 da
atmospheric pressure chemical ionization APCI	MH^+ $(M-H)^-$	M, Q	polar and some non-polar organic compounds, $\leq$ *ca.* 1000 da
fast atom/ion bombardment FAB/FIB/LSIMS	MH^+ $(M-H)^-$	M, Q	polar organics, proteins, organometallics, $\leq$ *ca.* 10 000 da (but depends on *m/z* range of MS)
field desorption/ ionization FD/FI	MH^+ $(M-H)^-$	M, (Q)	non-polar and some polar organics, inc. synthetic polymers, $\leq$ *ca.* 10 000 da (but depends on *m/z* range of MS)
thermospray TSP	MH^+, MNH_4^+ $(M-H)^-$	M, Q	polar and some non-polar organic compounds, $\leq$ *ca.* 1000 da
matrix assisted laser desorption ionization MALDI	MH^+ $(M-H)^-$	TOF	polar and some non-polar biopolymers, synthetic polymers *ca.* 200 000 da and higher

[a] M = molecular weight.
[b] M = magnet; Q = quadrupole; TOF = time-of-flight.

of each ionization method, but also to be able to choose the most appropriate method for the instrumentation and task in hand.

The ionization method has to be available and compatible with the mass spectrometer being used, in addition to being able to deal with the samples, and Table 1.1 lists the commonly used ionization methods, together with the type of sample information generated, the complementary mass spectrometer(s), and the general samples classes to which these ionization methods can be applied.[16]

Out of all the ionization methods listed, electron impact (EI; see Chapter 3) is the one that produces molecular ions and generally fragment ions as well, while all the others are termed '**soft ionization**' methods and generate quasimolecular ions. Electrospray is the only method that gives rise to multiply charged ions to any great extent, the other methods produce singly charged species and so the masses can be read directly from the m/z scale.

To summarize Table 1.1, if one has a time-of-flight mass spectrometer, most probably the sole ionization method available on the instrument will be matrix assisted laser desorption ionization (MALDI); if the mass spectrometer has a quadrupole mass analyser, then EI, chemical ionization (CI), electrospray (ES), atmospheric pressure chemical ionization (APCI), fast atom (or ion) bombardment (FA/IB), or thermospray (TSP) are compatible; if a magnetic sector mass spectrometer is available, it could be used with any of the ionization methods mentioned for quadrupole analysers, in addition to field desorption (FD) and field ionization (FI).

Which Ionization Methods are Appropriate for Different Sample Classes?

There is usually a choice of ionization method for any particular sample type, and the choice may be influenced by the mass spectrometer available, and any separation method that may be necessary. With this purpose in mind, Table 1.2, with the emphasis placed firmly on the type of sample, has been constructed. For each class of compounds, a list of ionization methods that may be employed gainfully in their analysis is given. For each ionization method listed, the different types of compatible chromatographic interfaces are presented, together with the possible types of mass spectrometers that could be used.

For example, if proteins are being analysed, electrospray ionization (on either a quadrupole or a magnetic sector mass spectrometer), fast atom bombardment (on either a quadrupole or a magnetic sector mass spectrometer), or matrix assisted laser desorption (on a time-of-flight mass spectrometer) could all be used to good advantage. If electrospray is chosen, the samples could be introduced directly into the ionization source (if the samples are sufficiently pure), or *via* an LC or CE interface, with on-line separation. On the other hand, if MALDI is employed, then the samples will be analysed (on the majority of commercial instruments) without prior separation.

At the other extreme end of the sample scale, it can be seen that GC is interfaced only to electron impact or chemical ionization, on either a quadrupole or a magnetic sector mass spectrometer.

---

[16] M. E. Rose and R. A. W. Johnstone, *Mass Spectrometry for Chemists and Biochemists*, Cambridge University Press, Cambridge, UK, 1982.
This book is recommended in general for further, more detailed reading.

Table 1.2 *Sample types compared with chromatographic interfaces, ionization methods and mass spectrometers*

Sample classes	Chromatographic interfaces	Ionization methods	Mass spectrometer[a]
proteins, peptides, oligonucleotides, oligosaccharides	LC, CE	ES	M, Q, TOF
	low flow LC, CE	FAB/FIB/LSIMS	M, Q
	none	MALDI	TOF
polar organic compounds	GC	CI	M, Q
	LC, CE	ES	M, Q
	LC	APCI	M, Q
	low flow LC, CE	FAB/FIB/LSIMS	M, Q
	none	FD/FI	M, (Q)
	LC	TSP	M, Q
non-polar organic compounds	GC, particle beam LC	EI	M, Q
	GC	CI	M, Q
	none	FD/FI	M, (Q)
synthetic polymers	none	FD/FI	M, (Q)
	none	MALDI	TOF

[a] M = magnet; Q = quadrupole; TOF = time-of-flight.

A Comparison of Liquid Chromatography–Mass Spectrometry Methods

Probably the chromatographic area with the highest number of possible options is LC–MS, and it is for this reason that Table 1.3 has been compiled. LC–MS has been interfaced with success to the following ionization methods: electron impact, electrospray, atmospheric pressure chemical ionization, fast atom/ion bombardment, and thermospray. There are advantages and disadvantages for all of these methods and so it is important to find the best technique for analysing the samples at hand.

The most appropriate ionization method for the type of sample must be considered, and then the LC requirements checked to see if there are any incompatibilities such as flow rate range, the necessary presence of any additives (*e.g.* buffers and matrices) vital to the ionization method and their impact on the chromatographic resolution, and whether the permissible solvents and buffers complement the chromatography required.

In general LC–MS systems are *not recommended* for use with inorganic mineral acids, involatile buffers (including phosphates and perchlorates, where there is a danger of explosion), and high levels (> 100 mM) of any additive. Aside from these exceptions, most LC–MS systems are compatible with a wide range of aqueous and organic solvents and mixtures thereof, and also volatile buffers such as ammonium acetate and ammonium hydrogen carbonate, and additives including formic, acetic, and trifluoroacetic (< 0.1% *v/v*)

Table 1.3 *Comparison of LC–MS methods*

Ionization source	Interface required	Optimum flow rates (min – max)	Notes
EI	particle beam	$0.5 - 2$ mL min^{-1}	higher flow rates only with a low aqueous content in the mobile phase
ES	ES probe	30 nL min^{-1} $- 1$ mL min^{-1}	wide range of solvents acceptable – probably the most universal technique
APCI	APCI probe	$0.2 - 2$ mL min^{-1}	wide range of solvents acceptable
FAB/FIB/LSIMS	continuous flow FAB probe	$1 - 10$ μL min^{-1}	need a matrix present all the time – can be present throughout the chromatographic run or added post-column
	thin layer chromatography (TLC) probe		TLC plate must be sprayed with a matrix
TSP	TSP probe	$0.5 - 2$ mL min^{-1}	wide range of solvents acceptable – need either an electrolyte present, *e.g.* NH$_4$OAc, or discharge electrode

acids, and bases of the trialkylamine and aqueous ammonia type. As usual, the manufacturer's instruction manual should be consulted before trying novel systems, as damage to the mass spectrometer should be avoided.

Sample Analysis, Data Acquisition and Spectral Interpretation

After deciding upon the most appropriate chromatography if any, the ionization method and the type of mass spectrometer, then the data acquisition can be initiated if the type of analysis, *i.e.* full scan, SIR, or one of the numerous MS–MS methods, has been specified beforehand.

If a full scanning acquisition has been compiled, then the spectrum needs to be studied to reap as much information as possible about the sample. If there are many background or impurity ions present resulting from the solvents, buffers, additives, or column bleed, then it may be worthwhile to subtract *carefully* one or more background scans away from the sample-related scans.

The **molecular weight** may be apparent, either from a molecular ion (as in electron impact ionization, see Chapter 3 for a more detailed discussion) or

Nominal mass loss occurring (mass units)	Functional group possibly associated with mass loss
-15	$-CH_3$
-16	$-O$ $-NH_2$
-17	$-OH$ $-NH_3$
-18	$-H_2O$
-28	$-CO$ $-C_2H_4$
-29	$-CHO$ $-C_2H_5$
-31	$-OCH_3$
-32	$-CH_3OH$ $-S$
-42	$-CH_2CO$ $-C_3H_6$
-43	$-C_3H_7$ $-CH_3CO$
-44	$-CO_2$
-45	$-CO_2H$ $-OC_2H_5$
-60	$-CH_3CO_2H$
-80	$-SO_3$
-162	$-$ hexose
-176	$-$ glucuronic acid

Figure 1.12 *Common fragmentation losses*

from protonated or deprotonated, singly or multiply charged ions (as with the 'soft' ionization methods). With these 'soft' ionization methods, some samples are prone to **adduct formation** and these ions can often help, but occasionally confuse, the molecular weight diagnosis. For example, weak sodium $(MNa)^+$ and potassium $(MK)^+$ adducts are often detected as the $(M + 23)^+$ and $(M + 39)^+$ positive ions respectively. In many cases a mixture of, for example, MH^+ and MNa^+ ions are present and the two ions together, with a difference of 22 da, reinforce the evidence for the molecular weight. If only one set of these quasimolecular ions is present, it is usually the MH^+ ions, but on the rare occasions when the sole quasimolecular ions relate to MNa^+ adducts, an incorrect molecular weight may have been assumed. It is necessary to study all the information in the spectrum, and not simply the ions that relate to the compound expected, or those ions that are the most readily identified.

If the sample is an organic compound with a molecular weight which is an odd number less than *ca.* 1000 da, this implies that there is an odd number of **nitrogen atoms** present, *e.g.* 1, 3, 5, *etc.* Conversely, if the molecular weight is an even number, then the compound contains an even number of nitrogen atoms, *e.g.* 0, 2, 4, *etc.*

If the sample is of a high molecular weight such as a biomolecule, and electrospray has been used for the analysis, then a series of multiply charged ions will have been detected and the molecular weight can be determined from these ions either manually or with software interpretation programs.

Having ascertained the molecular or quasimolecular ions, then it is useful to check the pattern of the **isotopes**. As discussed earlier in this Chapter, chlorine, bromine and certain transition metal atoms have more than one isotope of significant intensity and hence produce diagnostic isotopic 'fingerprints'.

Finally, if the sample has been analysed under conditions that enhance **fragmentation**, than it may well be possible to deduce structural information from the fragment ions in the spectrum.[2,17] If the structure of the compound under investigation is known then possible cleavage sites can be presumed and the *m/z* values of the theoretical fragment ions calculated. If the sample is completely unknown then *m/z* differences between the molecular related ions and any fragment ions in the spectrum should be calculated and some structural inferences can often be made. For example, a loss of 18 da often implies dehydration. Retro-Diels–Alder cycloadditions are possible. A number of frequently observed fragmentations have been tabulated and are displayed in Figure 1.12, but this is by no means an exhaustive list. The fragmentation behaviour of peptides, including both backbone and side-chain cleavages, has been well documented and a standard nomenclature exists.[18,19]

[17] F. W. McLafferty and F. Turecek, *Interpretation of Mass Spectra*, University Science Books, CA, USA, 4th edn, 1993.

[18] P. Roepstorff and J. Fohlman, *Biomed. Mass Spectrom.*, 1984, **11**, 601.

[19] R. S. Johnson and K. Biemann, *Biomed. Environ. Mass Spectrom.*, 1989, **18**, 945.

CHAPTER 2

Atmospheric Pressure Ionization Techniques – Electrospray Ionization and Atmospheric Pressure Chemical Ionization

1 What Type of Compounds can be Analysed by Atmospheric Pressure Ionization Techniques?

Atmospheric pressure ionization (API) encompasses two quite different ionization methods, electrospray[1] (ES) and atmospheric pressure chemical ionization (APCI).

High molecular mass samples (> *ca.* 1000 da) such as proteins,[2] peptides,[3] and oligonucleotides[4] are ideally suited to electrospray ionization. Electrospray produces multiply charged ions from such molecules, which often have molecular masses far in excess of the m/z range of the mass spectrometer. From these multiply charged ions the molecular mass of the sample can be determined readily.

For many lower molecular mass samples, if the sample has some degree of polarity and is in solution, both electrospray and atmospheric pressure chemical ionization should produce either protonated or deprotonated singly charged molecular ions, depending on whether positive or negative ionization respectively is used. In most cases there are no set rules for deciding which technique to use, although if the study is one which demands the highest level of sensitivity, *e.g.* a quantification assay, then it is worthwhile analysing the sample with both ionization techniques and making the final decision based on these practical results.

Classes of compounds analysed routinely by electrospray, in addition to the high molecular mass ones mentioned above, include peptides and protein

---

[1] J. Fenn, *J. Phys. Chem.*, 1984, **88**, 4451.
[2] B. N. Green, *Biochem. J.*, 1992, **284**, 603.
[3] P. A. Schinder, A. Van Dorsselaer and A. M. Falick, *Anal. Biochem.*, 1993, **213**, 256.
[4] A. Deroussent, J.-P. Le Caer, J. Rossier and A. Gouyette, *Rapid Commun. Mass Spectrom.*, 1995, **9**, 1.

digests,[5] sugars,[6] drugs and their metabolites,[7] phospholipids[8] and indeed most other organic compounds.

Classes of compounds analysed routinely by atmospheric pressure chemical ionization include steroids, pesticides,[9] drugs and their metabolites,[10] surfactants[11] and again, most other organic compounds.

2 Electrospray Ionization

The Principles of Electrospray Ionization

The first combined electrospray–mass spectrometry data were published in 1984[1] and since then the technique has been widely employed to analyse a variety of polar molecules ranging from < 100 da up to and above 200 000 da in molecular mass.

Electrospray ionization is implemented as an atmospheric pressure ionization technique on quadrupole and magnetic sector mass spectrometers (and now also on time-of-flight instruments), and as with all the API techniques, the formation of ions takes place outside the vacuum system of the mass spectrometer (Figure 2.1). Electrospray operates by an 'ion evaporation' process,[12] whereby ions are emitted from a droplet into the gas phase.

In principle, the sample, in solution, is introduced into the ionization source through a stainless steel capillary (75–100 μm internal diameter) contained within a mass spectrometric probe, at a flow rate of between 1 μL min^{-1} and 1 mL min^{-1}, but more typically in the region 5–300 μL min^{-1}. The solution used, which can range from 100% organic to 100% aqueous, is varied to suit the sample under investigation; however, a typical set-up would be 1:1 (*v/v*) water/acetonitrile or methanol. The concentration of sample needed to produce a full spectrum varies according to the sample's amenability to electrospray ionization and to the type of instrument being used, but typically would be in the region of 1–20 pmol μL^{-1} for higher molecular mass samples such as peptides, proteins, and oligonucleotides, and 1–50 ng μL^{-1} for samples up to 1000 da molecular mass.

A voltage of 3 or 4 kV is applied to the tip of the capillary once the probe is present in the source of the mass spectrometer, and as a consequence of this strong electric field, the sample solution emerging from the capillary is dispersed into an aerosol of highly charged droplets. This electrospray process

[5] K. F. Medzihradszky, D. A. Maltby, S. C. Hall, C. A. Settineri and A. L. Burlingame, *J. Am. Soc. Mass Spectrom.*, 1994, **5**, 350.

[6] J. Peter-Katalinić, A. E. Ashcroft, B. N. Green, F. G. Hamisch, Y. Nakahara, H. Iliyma and T. Ogawa, *Organic Mass Spectrom.*, 1994, **29**, 747.

[7] J. N. Robson, S. Draper and K. Tennant, *Anal. Proc.*, 1994, **31**, 159.

[8] P. Michelsen, B. Jergil and G. Odham, *Rapid Commun. Mass Spectrom.*, 1995, **9**, 1109.

[9] S. Bajic, D. R. Doerge, S. Lowes and S. Preece, *Am. Lab.*, 1993, **25**, 40.

[10] R. J. McCracken, W. J. Blanchflower, S. A. Haggan and D. G. Kennedy, *Analyst*, 1995, **120**, 1763.

[11] S. D. Scullion, M. R. Clench, M. Cooke and A. E. Ashcroft, *J. Chromatogr.*, 1996, **733**, 207.

[12] J. V. Iribane and B. A. Thomas, *J. Chem. Phys.*, 1976, **64**, 2287.

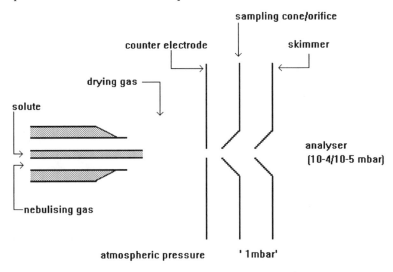

Figure 2.1 *Schematic diagram of an electrospray ionization source*

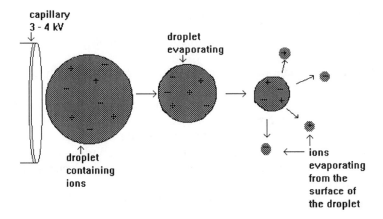

Figure 2.2 *The electrospray ionization process*

is aided by a co-axial nebulizing gas also flowing down the probe, around the outside of the capillary. This gas, usually nitrogen, helps to direct the spray emerging from the capillary tip. The charged droplets diminish in size by evaporation, assisted by a flow of warm nitrogen gas known as the drying gas, which passes across the front of the source. Eventually charged sample ions, free from solvent, are released from the droplets (Figure 2.2). Some of these ions pass through a sampling cone or orifice into a pumped intermediate vacuum region (*ca.* 1 mbar pressure), and then through a small hole, the skimmer, into the analyser of the mass spectrometer where their *m/z* ratios are measured. The analyser is under high vacuum. The skimmer acts as a momentum separator and the heavier sample ions pass through, while the

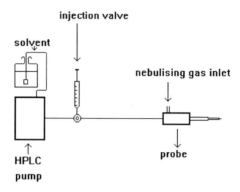

Figure 2.3 *Setting up electrospray mass spectrometry*

lighter solvent and gas molecules are pumped away in the intermediate vacuum stage.

Practical Operation of Electrospray Ionization

Essential Requirements for Operation

The following are prerequisites for electrospray operation:

- either a quadrupole or a magnetic sector (or more recently a time-of-flight) mass spectrometer;
- an electrospray source, probe, and any associated extra vacuum pumps and related vacuum lines, which should all be assembled and fitted to the mass spectrometer according to the manufacturer's manual;
- a supply of high purity nitrogen for use as a drying and nebulizing gas;
- an HPLC pump capable of delivering a pulse-free solvent flow at the required flow rate (see Section 4);
- degassed HPLC grade solvents and reagents (see Section 4);
- HPLC accessories, including syringe, injection valve (unless a syringe pump is being used), and associated ferrules, fittings, peek tubing, stainless steel capillary tubing;
- sample vials and pipettes for preparing sample solutions and dilutions.

The outlet of the HPLC pump should be connected via an injection valve (unless a syringe pump is being used) to the electrospray probe (Figure 2.3). For initial experiments, the use of an HPLC column should be avoided as the optimization procedure is far less complicated when simply injecting solutes directly to the ionization source.

Setting up Electrospray Mass Spectrometry

It is of paramount importance that the electrospray ion beam is stable before any sample analyses are undertaken, and several parameters play an important rôle in the practical operation of electrospray ionization.

Before inserting the electrospray probe into the ionization source of the mass spectrometer, a number of criteria should be checked. First, the solvent flow through the probe should be observed without either the nebulizing gas or the high voltage applied. The flow rate should lie in the range recommended for both the HPLC pump and the mass spectrometer, and it should be checked by timing a known volume as the solvent emerges from the capillary. This can be accomplished practically by using the barrel of a 50 or 100 μL syringe, or a disposable micro-pipette. An incorrect flow rate could indicate a faulty HPLC solvent delivery pump, or that solvent is leaking from one of the connections.

Solvent degassing is of importance, especially when operating at low flow rates, and it is useful to make a mental note of the backing pressure on the HPLC pump daily so that any solvent-related discrepancies can be seen immediately. A decrease in the backing pressure could indicate either an air bubble passing through the system, or solvent leakage. Conversely, an increase in the backing pressure may imply a blockage in the solvent delivery system, or the injection valve, or the capillary in the electrospray probe.

The source temperature is of importance, as too high a temperature will cause the solvent to start evaporating before it reaches the tip of the capillary, and too low a temperature will allow excess solvent to accumulate in the source. The temperature of the source depends on the composition and flow rate of the mobile phase, and again the manufacturer's instructions should be consulted to verify the optimum values attainable. Typical values for a 1:1 acetonitrile/water (v/v) mixture would be 70 °C for a 10 μL min^{-1} flow rate, and up to 180 °C for a 1 mL min^{-1} flow. Different solvents are discussed in more detail in Section 4.

The probe tip should be aligned such that the solute/solvent transporting capillary is concentric with respect to the nebulizing gas capillary, and that the former protrudes *ca.* 0.5 mm, or whatever is recommended for the probe in use.

After having verified these parameters, the nebulizing and drying gases can be switched on, the probe can be inserted into the mass spectrometer, and the mass spectrometric voltages all switched into operation.

The user should decide whether positive or negative ions are to be detected, and this choice depends very much on the type of sample to be analysed. In general, acidic compounds produce better results with negative ion detection, while basic samples respond more to positive ion detection. The presence of the ion beam should be confirmed, and this can be verified quickly by looking for solvent related ions. For example, if acetonitrile is present in the mobile phase, then MH$^+$ ions at *m/z* 42 should be present in positive ionization electrospray.

Having ascertained that the electrospray process is operating, it is now preferable to inject a sample of known molecular mass (which has been analysed successfully on other occasions) in order to check the electrospray performance. This is because the optimal tuning conditions for solvent-related ions do not always correspond to the optimum parameters required for sample analysis. Any known sample can be used for this purpose, but it is always preferable to choose a compound similar to those which will be analysed subsequently, *e.g.* horse heart myoglobin (MW 16 951.49 da; see Appendix 2) is suitable for tuning in either positive or negative mode prior to protein analyses, and erythromycin (MW 733.46 da) can be used before running low molecular mass samples in the positive ion mode. Suitable ions should be selected and monitored for ion beam stability, as an unstable, vastly fluctuating signal is not going to produce good quality, reproducible data. For horse heart myoglobin, the positive ions at *m/z* 1060.48, which in fact carry 16 charges, are suitable if using a slightly acidified aqueous/organic solvent and positive ion detection, while for erythromycin, the protonated molecular ions at *m/z* 734.47 are generally quite strong. It should be remembered that if using a low flow rate, say less than 30 μL min^{-1}, the sample may take a minute or so to reach the source, depending on the length and diameter of the capillary used between the HPLC pump and the electrospray probe. The sample should be dissolved in a solution, which is the same as, or compatible with, the mobile phase.

Now with sample ions present, the exact position of the probe can be optimized. A probe that is pushed too far into the source will cause electrical breakdown and hence source instability; a probe too far out of the source will not give the best sensitivity. By this time, with a reasonably steady ion beam, fine tuning of the voltages on the capillary and source lenses, and the gas flows, can be undertaken to ensure optimum results.

The cleanliness of the source should be maintained at all times as this will doubtless affect overall electrospray sensitivity and probably ion beam stability too. A daily check measuring the intensity of a known sample ion injected at a known concentration, either by data acquisition or simply using an oscilloscope if one is present, or a data system display, will take only a few minutes and provide confidence in the system.

If sample ions cannot be detected, then all of the procedures described here should be re-checked.

The Analysis of 'Low' Molecular Weight, Singly Charged Samples (up to *ca.* 1000 da)

Molecular Weight Determination

For samples with molecular masses up to *ca.* 1000 da, the ions produced from electrospray tend to be similar to those formed by other soft ionization techniques, *i.e.* a protonated molecular ion (MH^+) in positive ionization mode for basic compounds such as amines, or a deprotonated molecular ion $(M–H)^-$ in negative ionization mode for acidic compounds and compounds containing other electronegative groups. Some adducting may be observed also, for example, ammonium (MNH_4^+, M + 18), sodium (MNa^+, M + 23) or potassium (MK^+, M + 39) in positive ionization electrospray, or sodium $[(M–H + Na)^-$, M + 22] or formate $[(M–H + HCO_2)^-$, M + 44] in negative ionization electrospray. The degree of adducting depends on the sample, its preparation prior to analysis, its affinity for anionization or cationization, and the composition of the mobile phase and associated buffers. In general, few fragment ions are observed under standard electrospray operating conditions; fragmentation is usually achieved by some form of collision inducement (Chapter 2, Section 2). Electrospray ionization is therefore a good technique for producing molecular mass information on a wide range of low molecular mass samples including drugs and their metabolites, pesticides, small peptides, organometallics, and sugars.

The positive ionization spectrum of the macrocycle erythromycin is shown (Figure 2.4), where the predominant ion at m/z 734.5 corresponds to the protonated molecular ion, thus confirming the theoretical molecular weight of 733.46. This spectrum was obtained by a 10 μL injection of the sample dissolved in 1:1 (v/v) water/acetonitrile at a concentration of 10 ng μL^{-1} into a mobile phase of the same solvent flowing at 10 μL min^{-1} directly to the ionization source.

As always, mass accuracy is of importance and the mass spectrometer should be properly calibrated. Suitable calibrants for ES–MS analyses include polyethylene glycol mixtures for positive ionization analyses, and sodium iodide clusters for either positive or negative ionization studies (see Appendix 2).

Structural Elucidation

In addition to generating molecular weight information, fragmentation of the quasi-molecular ions can be achieved in order to generate structurally informative data. Fragmentation can be achieved in different ways, either by varying the electrospray source parameters to induce in-source fragmentation, or more specifically by using MS–MS techniques, if the type of mass spectrometer being used incorporates more than one analyser, *i.e.* is a tandem mass spectrometer.

The former method is achieved readily by increasing the voltage on the sampling cone or orifice, and is sometimes known as 'cone voltage fragmentation'. The sampling cone or orifice is used to extract the ions from the atmospheric pressure region of the source and transmit them into the intermediate vacuum region. From here, the ions accelerate towards the skimmer at the rear of this vacuum region, and thereafter enter into the high vacuum region of the mass spectrometer. By increasing the sampling cone voltage relative to the skimmer voltage, the ions are accelerated more quickly through this region, and collisions between the ions and any solvent vapour or gas present may now be of sufficient energy to cause the ions to fragment. Such low energy collisions often involve breaking of the more susceptible bonds such as the C–N and C–O ones.

The in-source collision induced decomposition spectrum of erythromycin is shown in Figure 2.5. The MH$^+$ ion is no longer the base peak in the spectrum, and two dominant ions at lower mass have appeared, *i.e.* m/z 576 and 158, the former resulting from loss of a sugar moiety from the molecule and the latter the amino-sugar group after cleavage.

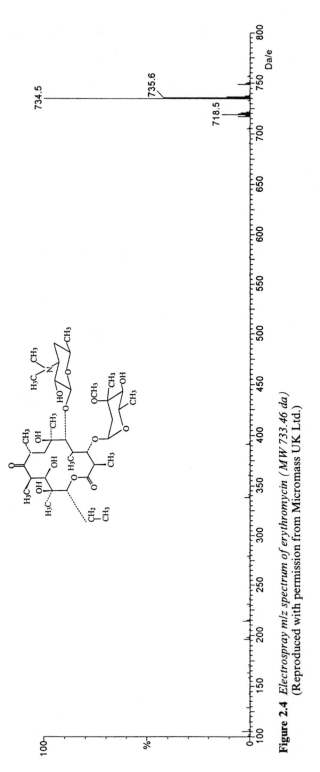

Figure 2.4 *Electrospray m/z spectrum of erythromycin (MW 733.46 da)*
(Reproduced with permission from Micromass UK Ltd.)

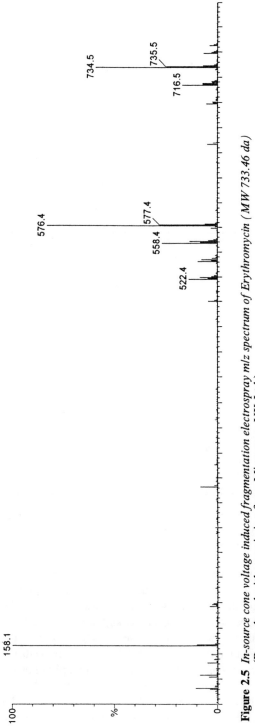

Figure 2.5 *In-source cone voltage induced fragmentation electrospray m/z spectrum of Erythromycin (MW 733.46 da)* (Reproduced with permission from Micromass UK Ltd.)

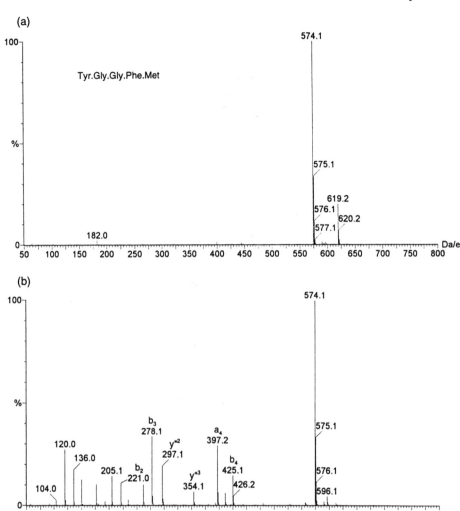

Figure 2.6 *Electrospray m/z spectra of methionine enkephalin (MW 573.22) obtained with: (a) a low cone voltage; (b) a high cone voltage*
(Reproduced with permission from Micromass UK Ltd.)

Another example of the use of in-source induced fragmentation is in the area of peptide analysis, where amino acid backbone fragmentations can frequently be used to provide structural, or in this case sequencing, information. The positive ionization electrospray spectra of methionine enkephalin (MW 573.22 da), obtained first with a low and then with a high cone voltage from a 10 μL injection of a solution of methionine enkephalin in 1:1 (v/v) acetonitrile/0.1% aqueous formic acid at a concentration of 10 ng μL^{-1} into a mobile phase of 1:1 (v/v) acetonitrile/water flowing at 10 μL min^{-1} to the source are shown in Figure 2.6. The first spectrum confirms the expected molecular mass of the sample by the presence of a dominant MH$^+$ ion at *m/z* 574, while the second spectrum shows quite extensive fragmentation. The amino acid backbone fragment ions have been labelled as the a, b, or y" ions on the high cone voltage spectrum, using standard Roepstorff[13] and Biemann[14] nomenclature.

This type of fragmentation is straightforward to set up, and requires the consumption of very little extra sample. It should be remembered, however, that not only does the sample of interest fragment, but also do any other components and any residual solvent-related ions present simultaneously in the source. Therefore this method, although informative and often invaluable, is not a specific means of inducing fragmentation, and in order to achieve the ultimate specificity, a mass spectrometer with multiple analysers must be used to obtain MS–MS spectra by the generation of fragment ions from a specific species, and then by monitoring these fragment ions, the precursor ions, or constant neutral losses. MS–MS is also the method of choice for oligonucleotide sequencing.[15]

One other observation which may be made with the implementation of higher cone or orifice voltages is that the relative intensities of any adduct ions may well increase relative to the protonated or deprotonated molecular ions.

Accurate mass measurements are possible under electrospray conditions, and polyethylene glycols have been used as reference compounds.[16]

The Analysis of 'High' Molecular Weight, Multiply Charged Samples

Higher mass samples such as proteins,[2,17] glycoproteins,[2,17] oligosaccharides and oligonucleotides[18] generally produce multiply charged ions under electrospray ionization conditions. These ions can be identified as a series that spans the m/z scale and contains a number of multiply charged ions of different charge states, each one differing from adjacent members of the series by one charge. The multiply charged ions are of the form $(M + nH)^{n+}$ or $(M - nH)^{n-}$ where n equals the number of charges on an ion, and as the mass spectrometer detects ions according to their m/z ratio, these ions tend to appear (fortuitously) in the range m/z 300 to 4000. For example, if a protein of molecular mass 30 000 da has 20 charges, the resulting ions will appear at m/z [(30 000 + 20 × 1.0078)/20], *i.e.* m/z 1501.01, if the atomic mass of hydrogen is taken as 1.0078. The greater the number of charges on a particular sample, the lower down the m/z scale the ions appear, and *vice versa*.

The phenomenon of routine multiple charging is exclusive to electrospray and makes it an extremely valuable technique in the biotechnology field, because mass spectrometers can analyse high molecular mass samples without

[13] P. Roepstorrf and J. Fohlman, *Biomed. Mass Spectrom.*, 1984, **11**, 601.

[14] R. S. Johnson and K. Biemann, *Biomed. Environ. Mass Spectrom.*, 1989, **18**, 945.

[15] J. Ni, S. C. Pomerantz, J. Rozenski, Y. Zhang and J. A. McCloskey, *Anal. Chem.*, 1996, **68**, 1989.

[16] A. N. Tyler, E. Clayton and B. N. Green, *Anal. Chem.*, 1996, **68**, 3561.

[17] 'Methods in Molecular Biology, Vol. 61: Protein and Peptide Analysis by Mass Spectrometry', ed, J. R. Chapman, Humana Press Inc., NJ, 1996.

[18] M. Greig and R. H. Griffey, *Rapid Commun. Mass Spectrom.*, 1995, **9**, 97.

any need for extending their mass range, and without any loss of sensitivity. In general, proteins and glycoproteins respond well to positive ionization electro-spray, while oligosaccharides and oligonucleotides prefer negative ionization electrospray. Often, a trace of acid (*e.g.* 0.01–0.1% formic acid) is added to enhance positive ionization and trace of base (*e.g.* 0.1–1% triethylamine) to enhance negative ionization.

The spectrum of horse heart myoglobin (molecular mass 16 951.49 da) (Figure 2.7) illustrates the type of charge distribution obtained from a single, pure protein. This sample was analysed by injection of 10 μL of a 10 pmol μL^{-1} solution in 1:1 (*v/v*) acetonitrile/0.1% aqueous formic acid into a mobile phase of 1:1 (*v/v*) acetonitrile/water flowing at 10 μL min^{-1} to the positive ionization electrospray source.

In order to calculate the molecular mass of the protein, two principles need to be applied. First, when analysing an unknown sample, then the charge state of any particular ion is also unknown, and so the assumption must be employed that any two adjacent members in the series of multiply charged ions differ by one charge. In the spectrum, two masses have been labelled: M_1, which has n charges, and M_2, which has $(n + 1)$ charges. Second, the general equation for calculating the m/z values of ions can be written as follows:

$$\frac{m}{z} = \frac{M + nH}{n} \tag{2.1}$$

where M = the molecular mass of the sample,
          n = the number of charges,
          H = the mass of one proton, and
          m/z = the mass-to charge ratio, or the apparent mass of the ion.

From Equation (2.1), it can be seen that:

$$M_1 = \frac{M + nH}{n} \tag{2.2}$$

and:

$$M_2 = \frac{M + (n+1)H}{(n+1)} \tag{2.3}$$

As the molecular mass (M) is the unknown quantity, this can now be removed from the simultaneous Equations (2.2) and (2.3) by rearranging them as follows:

$$nM_1 + H = (n+1)M_2 \tag{2.4}$$

and hence:

$$n = \frac{M_2 - H}{M_1 - M_2} \tag{2.5}$$

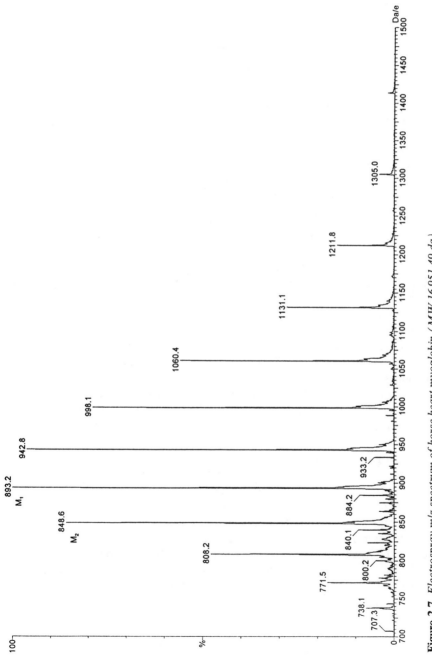

Figure 2.7 *Electrospray m/z spectrum of horse heart myoglobin (MW 16 951.49 da)*
(Reproduced with permission from Micromass UK Ltd.)

The values of M_1 and M_2 can be read directly from the spectrum, assuming that the instrument is accurately calibrated, and so the value of n can be calculated. Once the value of n is known, it can be substituted into Equation (2.2) to calculate the protein's molecular mass, M.

For example, using the spectrum of Figure 2.7, the value of n can be calculated using Equation (2.5);

$$n = \frac{848.6 - 1.0078}{893.2 - 848.6}$$
$$\cong 19$$

that is, the ions at m/z 893.2 have 19 positive charges.

Now inserting this information into Equation (2.2), together with the mass of a hydrogen atom (H), the molecular mass of horse heart myoglobin can be determined:

$$893.2 = \frac{M + (19 \times 1.0078)}{19}$$
$$M = 19 \times (893.2 - 1.0078)$$
$$= 16\,951.65 \quad \text{(theoretical mass} = 16\,951.49 \text{ da)}$$

Although this example was for a positive ionization analysis, similar principles can be applied equally well with negatively multiply charged series of ions, *i.e.* $(M - nH)^n$.

The molecular mass of the multiply charged analyte is usually calculated automatically by the data system of the mass spectrometer, which is of great help and importance when considering a multi-component mixture analysis that may well contain overlapping series of multiply charged ions with each component exhibiting completely different charge states.

Additional processing techniques can be used to generate molecular mass profiles from the m/z spectra, whereby all the components are transposed onto a true molecular mass scale from which their molecular masses can be read without any amendments or calculations. The molecular mass profile of horse heart myoglobin (Figure 2.8) illustrates this. For more complex glycoprotein mixtures, the production of a molecular mass profile provides an accurate summary of all components present, and the molecular mass profile of horse-radish peroxidase illustrates the amount of information which can be gleaned from complex samples (Figure 2.9). Presented in this format, the data are far easier to interpret than a complex m/z spectrum containing numerous series of multiply charged ions. This sample was analysed by injection of 10 µL of a 10 pmol µL^{-1} solution in 1:1 (v/v) acetonitrile/0.1% aqueous formic acid into a mobile phase of 1:1 (v/v) acetonitrile/water flowing at 10 µL min^{-1} to the positive ionization electrospray source.[19]

[19] B. N. Green and R. W. A. Oliver, *Biochem. Soc. Trans.*, 1991, **19** (4), 929.

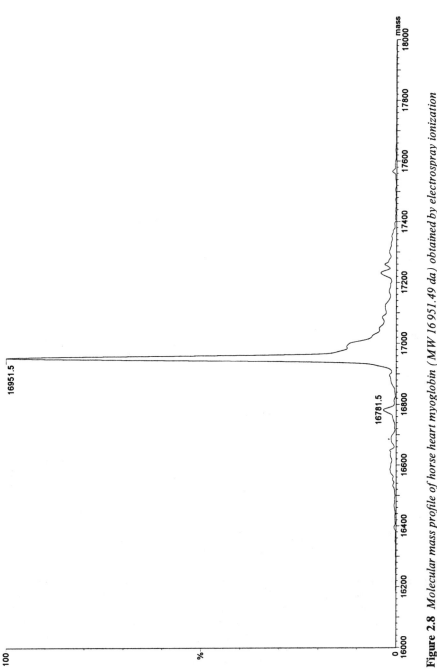

Figure 2.8 *Molecular mass profile of horse heart myoglobin (MW 16 951.49 da) obtained by electrospray ionization* (Reproduced with permission from Micromass UK Ltd.)

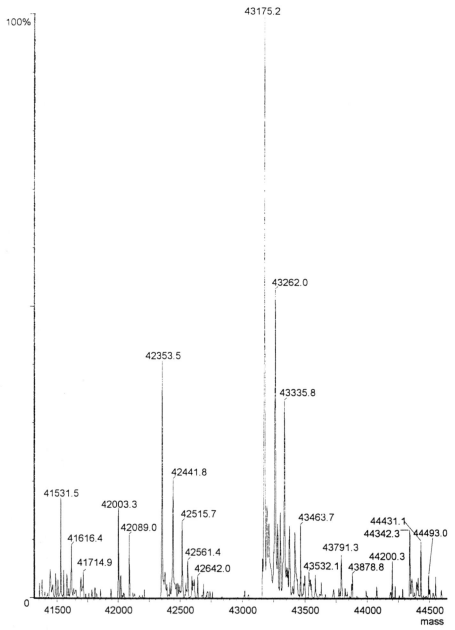

Figure 2.9 *Molecular mass profile of horseradish peroxidase (MW 43 261.8 da) obtained by electrospray ionization showing a mixture of components including the expected species together with a component without the C-terminal serine residue (MW 43 174.7 da), a component without the C-terminal serine residue but with an additional hexose (MW 43 336.8 da) and other components differing in their glycan residues*[19]
(Reproduced with permission from Micromass UK Ltd.)

Protein samples analysed under these conditions will be detected as highly charged ions representing the denatured state, most probably with any non-covalent bonds broken. Many elegant studies have been carried out in which proteins have been analysed in aqueous solution at various pHs in order to preserve non-covalent interactions.[20,21] Such studies have included protein–haem, protein–metal or enzyme–substrate complexes. These aqueous conditions are suitable for observing proteins in their conformed, or native, state and the resulting multiply charged ions generally have fewer charges and so occur at higher *m/z* values, even >*m/z* 3000 in some cases.[22]

In order to obtain accurate molecular mass measurements for those samples producing multiply charged ions, it is important to ensure that the mass spectrometer is properly calibrated. Although the low resolution spectrum of a low molecular mass sample will not be seriously affected by an error of 0.1 da, this same error on ions with say, 20 charges would result in a mass measurement inaccuracy of 20 times 0.1, *i.e.* 2 da, and could produce a misleading result. With a correctly calibrated mass spectrometer, mass inaccuracies of as little as 0.01% should be achievable with samples of molecular mass up to *ca.* 40 000 da. It is always advisable to calibrate the mass spectrometer with a high mass sample, if high mass analyses are to be performed. Horse heart myoglobin can be used in both positive and negative ionization modes quite satisfactorily (see Appendix 2).

These high molecular mass compounds do not tend to fragment readily, in contrast to the lower molecular mass compounds, although some MS–MS structural information has been obtained by monitoring the product ions from multiply charged ions with *n* as high as 19.[23]

Recent advances in electrospray include the development of the use of very low solvent flow rates (30 to 1000 nL min^{-1})[24,25] either as an LC column eluent or from a specially manufactured capillary sample vial. The former set up is ideal for coupling with nano-LC for the injection of small quantities of sample, while the miniature sample vials allow the continuous introduction of 1 or 2 μL of sample into the source over many minutes, providing the opportunity for reaction monitoring or for designing multiple mass spectrometric experiments.

Nanoflow electrospray not only consumes less solvent, but also less drying gas and nebulizing gas. The source temperature is held slightly above ambient, *e.g.* 30 °C, owing to the lower levels of solvent entering the mass spectrometer.

[20] C. V. Robinson, E. W. Chung, B. B. Kragelund, J. Knudsen, R. T. Aplin, F. M. Poulsen and C. M. Dobson, *J. Am. Chem. Soc.*, 1996, **36**, 8646.
[21] J. A. E. Kraunsoe, R. T. Aplin, B. N. Green and G. Lowe, *Fed. Eur. Biochem. Soc. Lett.*, 1996, **396**, 108.
[22] K. J. Light-Wahl, B. L. Schwartz and R. D. Smith, *J. Am. Chem. Soc.*, 1994, **116**, 5271.
[23] J. A. Loo, C. G. Edmonds and R. D. Smith, *Anal. Chem.*, 1993, **65**, 425.
[24] M. Wilm and M. Mann, *Anal. Chem.*, 1996, **68**, 1.
[25] M. T. Davis, D. C. Stahl, A. Hefta and T. D. Lee, *Anal. Chem.*, 1995, **67**, 4549.

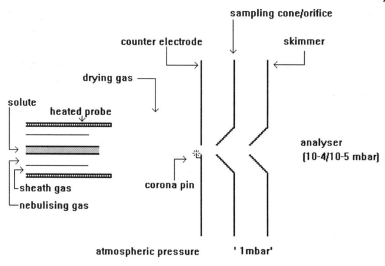

Figure 2.10 *Schematic diagram of an atmospheric pressure chemical ionization source*

3 Atmospheric Pressure Chemical Ionization

The Principles of Atmospheric Pressure Chemical Ionization

Atmospheric pressure chemical ionization (APCI)[26] has some similarities to electrospray ionization, in that for both cases ionization takes place at atmospheric, rather than reduced, pressure. The source is a modified electrospray source and so the extraction of ions into the mass spectrometer is much the same; the ionization process itself, however, is quite different, because with APCI the high voltage is not applied to the probe tip, and so nebulization and ionization are independent (Figure 2.10). The solute and solvent, flowing between 200 μL min⁻¹ and 2 mL min⁻¹, elute from a capillary which is surrounded by a co-axial flow of nebulizing gas (usually nitrogen) and sometimes a second, outer, coaxial gas known as the sheath gas (again usually nitrogen). The capillary and the gas(es) are contained in the triaxial probe, which is heated variably up to 700 °C according to the type of analyte under investigation. The source is held typically between 120 and 180 °C . The combination of nebulizer gas and heat convert the solvent flow into an aerosol, which then begins to evaporate rapidly. Inside the heated source is a corona discharge needle, which is held at a voltage of *ca.* 2.5 to 3 kV, and this is responsible for ionizing the solvent molecules. In the atmospheric pressure region around the corona pin, a combination of collisions and charge transfer reactions generates a chemical ionization reagent gas plasma. Any sample molecules which elute and pass through this region of

[26] E. Huang, T. Wachs, J. J. Conboy and J. D. Henion, *Anal. Chem.*, 1990, **62**, 713A.

solvent ions can be ionized by the transfer of a proton to produce $(M + H)^+$ or $(M - H)^-$ ions.

APCI tolerates a variety of solvents and buffers, ranging from 100% aqueous to 100% organic.

Practical Operation of Atmospheric Pressure Chemical Ionization

Essential Requirements for Operation

The following hardware is required:

- a quadrupole or a magnetic sector mass spectrometer;
- an APCI source, probe, and any associated extra vacuum pumps and related vacuum lines, which should all be assembled and fitted to the mass spectrometer according to the manufacturer's manual;
- a supply of high purity nitrogen gas for use as nebulizing, sheath and drying gas;
- an HPLC pump capable of delivering a pulse-free solvent flow at the required flow rate, (see Section 4);
- degassed HPLC grade solvents and reagents (see Section 4);
- HPLC accessories, including syringe, injection valve and associated ferrules, fittings, peek tubing, stainless steel capillary tubing;
- sample vials and pipettes for sample preparation.

The outlet of the HPLC pump should be connected via an injection valve to the APCI probe in the same way as for electrospray operation (Figure 2.3). For the initial optimization of the system, it is easier to operate without an HPLC column in-line and to inject solutes directly into the ionization source of the mass spectrometer.

Setting up APCI–MS

When setting up APCI, the solvent flow is first checked with the probe out of the mass spectrometer. A steady flow of solvent should emerge from the probe tip. Then, with the solvent flow **off** and the drying, sheath and nebulizer gases **on**, the probe is inserted into the source of the mass spectrometer. The temperature of the source should be set according to the manufacturer's instructions, and although it will depend on the precise flow rate and type of solvent being used, the typical temperature range is 120 to 180 °C. Now the temperature of the probe tip can be raised, say to 200 °C initially, and then only at this stage should the solvent flow be applied. A useful solvent to start with is 1:1 (*v/v*) acetonitrile/water at a flow rate of $0.5-1 \text{ mL min}^{-1}$.

The source voltages can be switched on, and a background spectrum of solvent ions should be seen once a sufficient voltage has been applied to initiate the corona discharge. The flow rate of the mobile phase is usually higher in APCI than in electrospray, and this means that not only does the sample take less time to reach the source, but that the length of time available for tuning sample-related ions is also significantly reduced. Often a more appropriate method of tuning is to acquire data through the data system, and monitor the most significant sample-related ion. The sample should be one whose APCI characteristics are known to the operator. Then the sample can be injected repeatedly, and in-between injections adjustments can be made to the various source parameters, including gas flows, the cone voltage, and the probe temperature, in a methodical and stepwise fashion. The increase or decrease in intensity of the ions under observation can be seen, and further adjustments made accordingly until the optimum performance has been achieved.

The Analysis of Samples

Unlike electrospray, APCI does not produce multiply charged ions and so is not appropriate for the analysis of high molecular mass samples such as proteins. APCI produces spectra dominated by either the $(M + H)^+$ or $(M - H)^-$ ion, depending on whether positive or negative ionization has been set, and so is ideal for the molecular mass determination of samples up to *ca.* 1200 da, as illustrated with the positive ionization spectrum of reserpine which shows a dominant ion at *m/z* 609, confirming the expected molecular mass of 608.27 da (Figure 2.11). This spectrum was obtained from a 10 μL injection of a 10 ng μL^{-1} solution of the sample in 1:1 (*v/v*) acetonitrile/water into a mobile phase of the same solvent flowing at 1 mL min^{-1} to the APCI source, with the source temperature set at 120 °C and the probe temperature set at 550 °C.

Although a high temperature is applied to the probe, most of the heat is used for solvent evaporation and heating the nebulizer gas, with the result that the thermal effect on the sample is much less than may well have been expected. However, with very labile compounds, the heated probe may still cause some fragmentation or decomposition.

In order to produce structurally informative fragment ions then the same methods exist as for electrospray, *i.e.* in-source cone voltage fragmentation, or MS–MS if a tandem mass spectrometer is available. The use of cone or orifice voltage fragmentation is illustrated for reserpine, which produces a variety of fragment ions (Figure 2.12).

4 Separation Methods Coupled to Atmospheric Pressure Ionization Techniques

Liquid Chromatography

HPLC Column Selection

The choice of HPLC column depends on the separation to be performed and on the HPLC solvent delivery systems available. As a general guide, the solvent flow rates of Table 2.1 are suggested for these different HPLC columns.

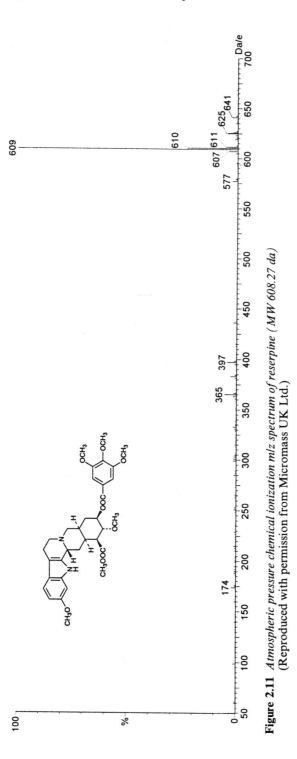

Figure 2.11 *Atmospheric pressure chemical ionization m/z spectrum of reserpine (MW 608.27 da)* (Reproduced with permission from Micromass UK Ltd.)

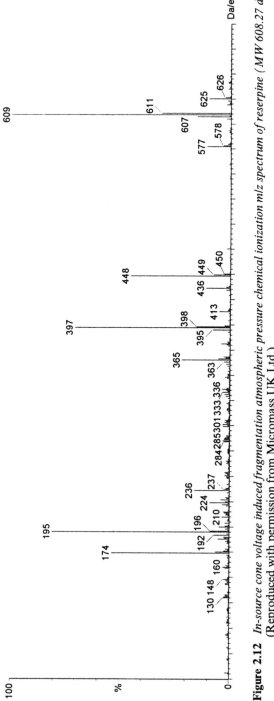

Figure 2.12 *In-source cone voltage induced fragmentation atmospheric pressure chemical ionization m/z spectrum of reserpine (MW 608.27 da)* (Reproduced with permission from Micromass UK Ltd.)

Table 2.1 *Suggested flow rates for HPLC columns*

HPLC column i.d. (mm)	Flow rate (min^{-1})
4.6	1.0 mL
2.1	200 μL
1.0	40–50 μL
0.3	5 μL
0.05	100 nL

The ionization technique required should also be taken into account when considering the HPLC column. For example, if a 4.6 mm i.d. HPLC column with a 1 mL min^{-1} flow rate is to be used, and electrospray is the preferred mode of ionization, then the ability of the electrospray source and probe to cope with high flow rates should be checked by referring to the manufacturer's instructions. If the probe and source are not suited to 1 mL min^{-1} flow rates, it could mean that the eluent needs to be split down after the HPLC column (see later in Section 4.1). Conversely, a column of less than 2.1 mm i.d. will probably not have a sufficiently high flow rate for APCI operation.

HPLC Solvent Delivery Pump Selection

If samples are being introduced into the ionization source directly, without prior separation, then the pump must be capable of delivering an isocratic pulse-free flow between 5 μL min^{-1} and 1 mL min^{-1} for electrospray operation, or between 200 μL min^{-1} and 2 mL min^{-1} for atmospheric pressure chemical ionization. If on-line LC–MS is required, then the choice of LC column often dictates the type of HPLC solvent delivery pump required.

Reciprocating pumps are popular and work under continuous operation because the pump reservoirs are continuously filled. They are generally used with HPLC columns of 4.6 or 2.1 mm internal diameter as the ability of this type of pump to perform gradient elutions at flow rates below 200 μL min^{-1} varies considerably, and should be verified before use at these low flow rates. Reciprocating pumps often have an associated autosampler for routine, unattended operation

Syringe pumps, with varying solvent capacity, usually offer gradient elution for flow rates up to 200 μL min^{-1} ; flows higher than this are often limited by time constraints as once the syringe pump is emptied, the flow of solvent stops until the pump is refilled. Syringe pumps are most commonly used in conjunction with HPLC columns ranging from 0.3 to 2.1 mm internal diameter. Syringe pumps can be adapted for use with HPLC columns of less than 0.3 mm internal diameter.[27]

For tuning purposes and constant infusion of a weak sample, or one which requires an unusual solvent system, an **infusion pump** can be used to best advantage.

[27] J. P. Chervet, M. Ursem and J. P. Salzmann, *Anal. Chem.*, 1996, **68**, 1507.

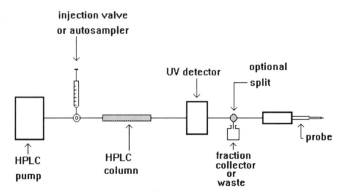

Figure 2.13 *Schematic diagram of a possible HPLC–UV–API–mass spectrometry config-uration*

HPLC–API–MS with In-line UV Detection

Frequently an HPLC method is developed with ultra-violet (UV) detection first, and in such cases it is often desirable to incorporate the UV detector into the solvent flow between the outlet of the HPLC column and the mass spectrometer, so that the two detectors are in simultaneous use during the same analysis. This allows direct comparison of the results from the one chromatographic separation and is often useful as the two detectors may exhibit different responses to the different components in the analyte. Many modern mass spectrometer data systems can also monitor the UV detector to facilitate this, and if this is the case, it should be remembered that when the UV detector is positioned before the mass spectrometer, there will be an offset of several seconds in the retention times of the eluting peaks on the UV output compared with the mass spectrometric output (Figure 2.13).

HPLC–ES–MS with Flow Splitting

For some electrospray analyses, if a 4.6 mm internal diameter HPLC column is used with a mobile phase flowing at 1 mL min^{-1}, the operator may choose to send only a portion of the eluent through to the mass spectrometer, in which case the eluent must be 'split' after the HPLC column. In such cases, just a fraction of the sample and mobile phase is directed to the electrospray ionization source, while the remainder may be sent to an alternative detector, a fraction collector, or simply to waste. This splitting may be performed for a variety of reasons: it could be to save valuable sample, or to keep the electrospray source clean for longer periods, especially when dealing with complex matrices such as those encountered with biological or environmental samples, or simply because the electrospray interface was not designed to carry high volumes of solvent.

Splitting devices are available commercially, or alternatively a split can be

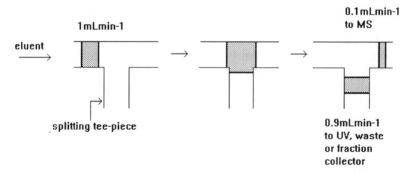

Figure 2.14 *Mobile phase flow splitting for electrospray operation*

set up quite simply by using a low volume T-piece, whereby the eluent from the column enters the T-piece, and the capillary leading to the electrospray probe exits the T-piece directly opposite to this. Another length of capillary is placed perpendicular to these two, in the third outlet of the T-piece, and this is varied in length to make a suitable split (Figure 2.14). The flow through the electrospray probe should be measured with a disposable micro-pipette, or by using the barrel of a 50 or 100 μL syringe.

A flow through the HPLC column of 1 mL min^{-1} would be split typically at a ratio of 1:9, whereby 0.1 mL min^{-1} is sent to the electrospray ionization source, and the remainder to a fraction collector or waste. Although it appears that a significant amount of analyte is not being analysed by the mass spectrometer, the detection limit appears to remain the same because the *concentration* of the analyte is unchanged.[28]

Solvent Selection

The type of solvent chosen for the mobile phase depends to a great degree on the type of sample under scrutiny, and the solvent used for dissolving the sample should be the same as, or at least totally miscible with, the initial mobile phase, otherwise sample precipitation may lead to capillary blockage. With on-line HPLC separations, thought must always be given to the impact of the HPLC mobile phase on the operation and performance of the mass spectrometer. Whatever the choice of solvents, they should always be of HPLC grade and degassed thoroughly before use.

The most common polar solvents used with API techniques are **water, methanol, acetonitrile, propan-2-ol** and mixtures thereof, with **1:1 (*v/v*) water/ acetonitrile** or **water/methanol** being used frequently when the sample is to be analysed by a discrete injection, or by constant infusion using a syringe pump.

[28] T. R. Covey, 'The Realities and Misconceptions of Electrospray Ionization and HPLC Flow Rates', Application Note, PE Sciex, Ontario, Canada.

Other, less used solvents include **ethanol, 2-methoxyethanol** and **dichloro-methane**, the latter showing success with some organometallics. **Tetrahydro-furan** has been used for size exclusion chromatography, but is not recommended for mass spectrometric work due to its flammability. It should be avoided completely if air is being used as either the nebulizer or the drying gas, and also if 'peek' tubing is in use.

Additives are often used in the mobile phase, whether an HPLC column is present or not, either to increase ionization by aiding the protonation or deprotonation of the sample, or to enhance the HPLC separation and resolution, if applicable.

For example, for positive ionization of proteins and peptides, as well as many other organic molecules, a small amount of acid is often added to the solute. **Formic acid** and **acetic acid** can both be used, generally at the 0.01 to 1% level. **Trifluoroacetic acid** is also used, as this somewhat stronger acid is popular for the separation of peptide mixtures arising from protein digests. If more than 0.1% of trifluoroacetic acid is present in the mobile phase, however, the sensitivity of the electrospray analysis may be significantly reduced. It should be remembered that the presence of acid aids the protonation of a basic sample in positive ionization, but will reduce the sensitivity of negative ionization operation quite considerably.

For negative ionization, **aqueous ammonia solution** (ammonium hydroxide) will promote the deprotonation of acidic samples, and is also useful for adjusting the pH of the mobile phase, if this is necessary. Oligonucleotides and oligosaccharides both respond well when this is added to the sample or in the mobile phase. **Triethylamine** is another basic additive which is appropriate for negative ionization analyses, but does protonate to produce an intense ion at m/z 102 in positive ionization operation.

Buffers can be used; the most suitable ones are those which are volatile and will therefore not precipitate onto the source, eventually reducing performance. **Ammonium acetate** is a common one, and is routinely used at concentrations of up to 100 mM. **Ammonium hydrogen carbonate** and **ammonium carbonate** have also been used. Non-volatile salts (*e.g.* **phosphates, perchlorates** and **borates**) and **inorganic acids** (*e.g.* **sulfuric** and **phosphoric acid**) should be avoided as these can lead to source contamination, and in the case of perchlorates, possibly explosions. Recent improvements in source designs mean that some sources are now beginning to cope with involatile buffers.

Examples of HPLC–API–MS

A Peptide Separation. An example of the on-line separation of the products from a protein digest mixture with electrospray ionization mass spectrometry is shown for bovine haemoglobin which was digested with trypsin and then separated using a 1 mm i.d. Aquapore RP-300 (Applied Biosystems Inc., Foster City, CA, USA) column with a mobile phase gradient of 5 to 95% (*v/v*) acetonitrile/0.01% aqueous trifluoroacetic acid over 45 min at 50 μL min^{-1} (Figure 2.15). The majority of the expected digest fragments were identified

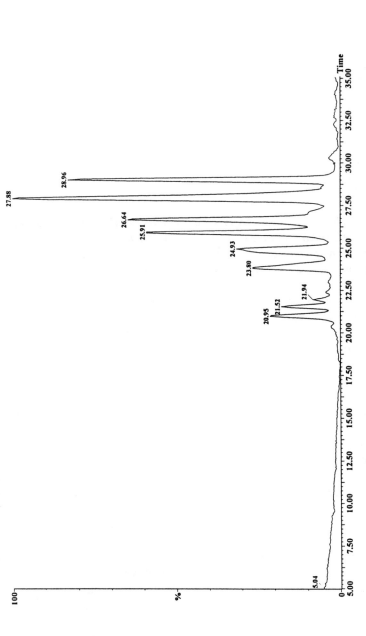

Figure 2.15 *On-line HPLC–MS separation of a tryptic digest mixture of bovine haemoglobin by electrospray operation. A gradient elution of 5 to 95% aqueous acetonitrile with 0.01% trifluoroacetic acid added was set at a flow rate of 50 μL min⁻¹ over 45 min. All of the eluent was directed to the electrospray ionization source. A 1 mm internal diameter HPLC column was used, and 5 μL of a nominally 20 pmol μL⁻¹ solution were injected*
(Reproduced with permission from Micromass UK Ltd.)

from their mass spectra, and could be used to search commercial protein libraries (*e.g.* the European Molecular Biology Library, EMBL) to confirm the identity of the original protein.

A Pesticide Separation. The separation of a mixture of phenyl ureas was achieved using a 4.6 mm i.d. C18 HPLC column with 60% (*v/v*) aqueous acetonitrile as the mobile phase at a flow rate of 1 mL min^{-1} (Figure 2.16). 50 ng of each of the phenyl ureas were injected, and analysed by on-line atmospheric pressure chemical ionization–mass spectrometry. Monolinuron, monuron, methabenzthiazuron, and diuron were identified from their positive ionization mass spectra. Herbicides can also be successfully analysed under the same conditions.[9,29]

Capillary Electrophoresis

Interfacing CE to ES–MS

A modified electrospray probe can be used to interface standard capillary electrophoresis (CE) equipment to the electrospray source of a mass spectrometer. The principles of capillary electrophoresis are little altered, but several considerations should be taken into account when coupling to electrospray. At all times the operator must be aware of the high voltages involved with capillary electrophoresis, and exercise due care and attention.

Two types of interface are most commonly used for this coupling: the triaxial or alternatively the liquid junction interface, and a recent review indicates that it is the former which is in more widespread use than the latter.[30]

The triaxial sheath flow interface involves taking the outlet electrode of the electrophoresis capillary directly into the ionization source of the mass spectrometer (Figure 2.17). It is important that the interface maintains the high separation efficiency of capillary electrophoresis, while taking into account the demands of the mass spectrometer.

If the recommended lowest flow rate into the electrospray source is 2 μL min^{-1}, then a solvent make-up flow is required because the flow emerging from the capillary electrophoresis system will be in the nL min^{-1} range. Within the triaxial probe, the separation capillary is surrounded by another capillary containing the solvent make-up flow, and then finally the nebulizing gas flows around these two capillaries. The solvent make-up flow is usually a 1:1 (*v/v*) mixture of methanol/water containing 0.1% formic or acetic acid for positive ionization electrospray, or 1:4 (*v/v*), methanol/propan-2-ol[31] for negative ionization electrospray, in both cases flowing at 5 to 10 μL min^{-1}. Acetonitrile is not recommended as this softens the polyimide coating of the

[29] D. R. Doerge and S. Bajic, 'Pesticide Analysis using APCI LC–MS', Application Note No. 207, VG Organic, Manchester, UK.
[30] M. W. F. Nielen, *J. Chromatogr.*, 1995, **712**, 269.
[31] R. F. Straub and R. D. Voyksner, *J. Am. Soc. Mass Spectrom.*, 1993, **4**, 578.

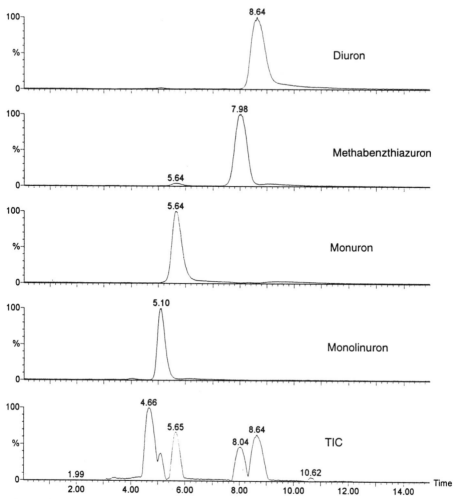

Figure 2.16 *On-line HPLC–MS separation of a phenyl urea pesticide mixture by on-line LC–APCI analysis. Diuron, methabenzthiazuron, monuron, and monolinuron have been highlighted*
(Reproduced with permission from Micromass UK Ltd.)

CE capillary. The solvent make-up flow also serves to make an electrical contact between the capillary electrophoresis buffer and the probe tip.

As may be expected, the concentricity and configuration of these different capillaries with respect to each other at the probe tip play a significant rôle in maintaining the separation efficiency, and the overall stability and sensitivity of the system. Typically the CE capillaries used are of the order 30 to 75 μm internal diameter, and the length needs to be sufficient to bridge the distance between the outlet of the capillary electrophoresis system and the source of the mass spectrometer. It is usual to operate with the minimum length of capillary, but as this may be up to 100 cm, it is still rather longer than the length used for

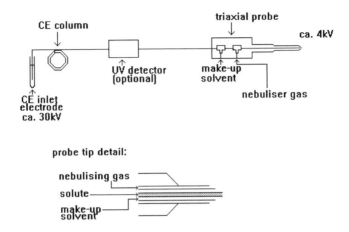

Figure 2.17 *Diagram of a triaxial sheath flow capillary electrophoresis – electrospray interface*

off-line CE analyses and so the separation takes a proportionally longer time. This is being addressed currently by at least one group.[32] The polyimide coating at the end of the separation capillary is flame-removed before use, and is set to protrude *ca.* 0.2 mm relative to the solvent make-up flow capillary, which in turn protrudes *ca.* 0.5 mm relative to the nebulizing gas capillary. Another important point is that the inlet and outlet (probe tip) electrodes of the capillary must be maintained at the same height to avoid any gravitational effects, and so the capillary electrophoresis unit is often placed on a stable, but adjustable trolley so that this parameter is accurately controlled.

CE–ES–MS Operation

For operation, the column is conditioned before use in the usual manner, *i.e.* consecutive rinses with sodium hydroxide solution, water, and the separation buffer to be used. The probe should be removed from the source if high concentrations of sodium hydroxide are being rinsed through, otherwise the source will become contaminated. It is then useful to make sure that the system is operating efficiently by placing the probe in the source, with the nebulizing and drying gases turned off, and pressure rinsing through a known sample, such as a small peptide, *e.g.* methionine enkephalin or bradykinin. Once the sample ions appear in the source, the high voltage of the capillary electrophoresis system can be applied, along with the nebulizing and bath gases, and the source conditions optimized as usual. It may be found that these gases need only to be used at very low flow rates. If there is no evidence for sample ions, then in addition to any of the points outlined previously under interfacing CE

---

[32] B. L. Krager, ASMS Fall Workshop 'Capillary Electrophoresis Mass Spectrometry', Virginia, USA, October 1995.

to ES–MS, it could be that either the capillary is blocked, or less drastically, that the sample level in the inlet vial is too low. When the operator is satisfied with the quality of the signal, the sample can be injected, either by a pressure or a voltage injection, again with the gases switched **off**, and then the buffer, capillary electrophoresis and mass spectrometric voltages, gases, and data acquisition initiated for the separation and detection of the analysis. The value of the capillary electrophoresis current should be noted at the beginning of a run, and if this current drops significantly, it is possible that there is an air bubble lodged in the separation capillary. This can usually be cleared by pressure rinsing with more buffer.

As with LC–ES–MS, the compatibility of the buffer with the mass spectrometer should be considered, and volatile buffers pose least problems. Ammonium acetate, taken to an appropriate pH by the addition of a volatile acid or base (*e.g.* acetic acid or aqueous ammonia solution) is a favourite starting point. Buffers such as borate and phosphate can crystallize in the source and eventually cause blockages and lead to electrical charging. Although non-volatile buffers are never recommended wholeheartedly, it should be remembered that the flow through the separation capillary is very low and any buffer present will be substantially diluted by the solvent make-up flow, and hence may be tolerated for limited periods of time.

The capillary electrophoresis–positive ionization electrospray separation of a mixture of peptides[33] is shown (Figure 2.18). The mixture consisted of 0.5 mg of each peptide in 10 mL of the buffer, and was analysed using a 5 s pressure injection followed by 30 kV separation on a fused silica capillary (75 μm i.d.; 375 μm o.d.; 90 cm length) with a buffer of aqueous ammonium acetate (15 mM) taken to pH 4 with 4:1 (*v*/*v*) acetic acid/acetonitrile. The make-up flow for this analysis was 1:1 (*v*/*v*) water/methanol at 10 μL min^{-1}, and the electrospray probe tip was held at 4 kV, thus maintaining a potential difference of $(30 - 4)$ = 26 kV between the inlet and outlet electrodes.

For this experiment, a UV detector was situated in-line, approximately 19 cm from the beginning of the separation capillary, which had a total length of 90 cm. The components elute on the UV electropherogram significantly earlier than on the mass spectrometric electropherogram: as might be expected by a factor of *ca.* five. It can be seen that an efficient separation has been achieved, and there is in fact very little peak broadening due to the mass spectrometric interface. The components were identified by their mass spectra as: bradykinin (MW 1059.56 da; retention time 13.16), angiotensin II (MW 1045.53 da; retention time 14.14), thyrotropin-releasing hormone (MW 362.17 da; retention time 16.03), bombesin (MW 1618.82 da; retention time 17.05), leucine enkephalin (MW 555.27 da; retention time 18.80), and methionine enkephalin (MW 573.22 da; retention time 19.08).

It should also be noted that although the initial concentration of sample was quite high, 50 ng μL^{-1}, only a very small injection volume was consumed. A 5s pressure injection can be estimated at consuming 30 nL of solution, or 1.5 ng

---

[33] H. J. Major and A. E. Ashcroft, *Rapid Commun. Mass Spectrom.*, 1996, **10**, 1421.

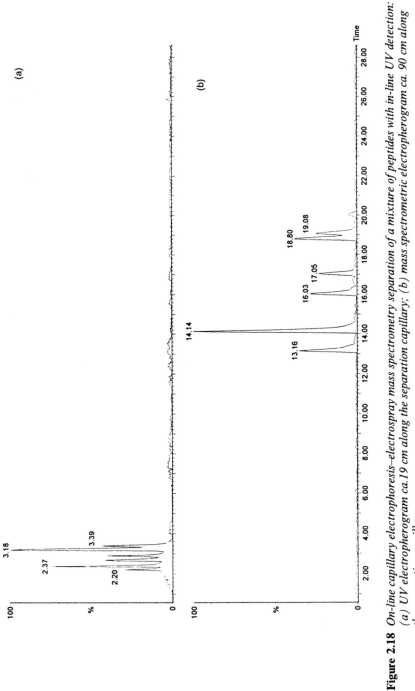

Figure 2.18 *On-line capillary electrophoresis–electrospray mass spectrometry separation of a mixture of peptides with in-line UV detection: (a) UV electropherogram ca.19 cm along the separation capillary; (b) mass spectrometric electropherogram ca. 90 cm along the separation capillary*
(Reproduced with permission from Micromass UK Ltd.)

of each component in this mixture. Thus this separation technique has the benefit of using very little sample, but does require a high sample concentration due to the minute amounts injected. This phenomenon has led to a great deal of effort being expended into devising means of multiple injections or of 'stacking' injections at the capillary inlet before applying the voltage to perform the separation. Transient isotachophoresis[30,34] has been used, and preconcentration cartridges whereby up to 2 μL of solute have been injected[35] and proved very successful.

[34] L. M. Benson, A. J. Tomlinson, J. M. Reid, D. L. Walker, M. M. Ames and S. Naylor, *J. High Resolut. Chromatogr.,* 1994, **16**, 151.

[35] A. J. Tomlinson, W. D. Bradock, L. M. Benson, R. P. Oda and S. Naylor, *J. Chromatog. B,* 1995, **669**, 67.

CHAPTER 3

Electron Impact and Chemical Ionization

1 What Type of Compounds can be Analysed by Electron Impact and Chemical Ionization Techniques?

The description of ionization methods in the pre-1980 mass spectrometric textbooks is restricted almost entirely to **electron impact (EI)** in the first instance, followed by **chemical ionization (CI)** for those samples which did not produce satisfactory EI data. Although the choice of ionization techniques in these earlier days was very limited, it should be remembered that substantial numbers of high quality spectra were generated in numerous application areas using complex, highly engineered mass spectrometers, requiring considerable dexterity to operate. Not only were spectra generated (and in the early days without the aid of a computer) but they were also interpreted to solve real-life problems. It is the author's personal belief that although we can now analyse a greater variety of compounds with a wider choice of ionization methods and chromatographic coupling in a much reduced time scale, we no longer need to understand fully every individual component of the mass spectrometer nor do we need to ponder over small fragment ions now that software is available for this purpose, and so perhaps over the years as our developments have so efficiently progressed, we are all culpable to some extent of taking for granted the results and the elaborate instruments responsible for generating these results.

Electron impact can be used for analysing a wide range of **volatile organic compounds** up to *ca.* 1000 da molecular weight including hydrocarbons, oils, alkaloids, steroids, pesticides, dioxins, flavours, fragrances, *etc.* EI is often used in conjunction with gas chromatography (GC) and usually if the sample can be analysed by GC, then it will be compatible with EI also. The two are an ideal match. EI is used in the positive ionization mode and tends to generate the $M^{+\bullet}$ of the sample (from which the molecular weight can be deduced directly) together with a significant number of fragment ions.

Chemical ionization is a soft ionization method that can be used in either the positive or negative mode and generally produces MH^+ or $(M-H)^-$ ions respectively with little evidence for fragment ions. This technique covers the same compound classes outlined for EI, and is commonly used when the molecule under investigation fragments so completely in EI that there is no evidence for the $M^{+\bullet}$ ions, or if knowledge of the molecular weight of the sample is of sole importance, or when good detection limits are required and so it is more beneficial to generate strong quasimolecular ions than divide the ion current between molecular and fragment ions.

EI and CI are used less frequently now due to the availability of other ionization methods, especially those more compatible with coupling directly to LC. Application areas such as pharmaceuticals, which may at one time have used GC–MS following derivatization of the relatively involatile drugs and metabolites, can now use LC–MS and avoid the derivatization step. However, some important application areas, notably the fields of dioxins,[1,2,3] pesticides,[4] and petroleum,[5] do use and need EI and CI, and they should certainly not be forgotten despite the more glamorous, newer ionization methods.

2 Electron Impact Ionization

The Principles of Electron Impact Ionization

The use of EI and CI in conjunction with magnetic sector and quadrupole mass spectrometers is described in this monograph.[6,7,8]

The heated EI source (Figure 3.1) operates under a high vacuum (*ca.* 10^{-5} to 10^{-6} mbar) into which samples are admitted as a vapour. The source contains a filament, usually made from tungsten, which is heated in order to emit electrons. These electrons are accelerated to the ionization chamber by a potential difference (the electron energy). The electrons then pass through a small aperture into the chamber and once there interact with the vaporized sample molecules. Additionally, the electron beam is collimated by a modest magnetic field generated by two small source magnets. The potential difference applied to accelerate the electrons is conventionally set to 70 eV (this is the difference in voltage between the filament and the ion chamber) although 20 eV is sufficient to ionize most sample molecules and still leave enough energy spare for fragmentations. Often the ions formed by initial fragmentation have sufficient energy to fragment further. Ionization of the sample molecules occurs by removal of an electron after collision with the electron beam, and as organic molecules are even-electron species, *i.e.* stable molecules in which all the electrons are paired, the process of removing one electron generates a

[1] J. R. Hass, M. Friesen and M. K. Hoffmann, *Org. Mass Spectrom.*, 1979, **14**, 9.

[2] M. L. Gross, T. Sun, P. A. Lyon, S. F. Wojinski, D. R. Hilker, A. E. Dupuy and R. G. Heath, *Anal. Chem.*, 1981, **53**, 1902.

[3] J. R. Chapman, G. A. Warburton, P. A. Ryan and D. Hazelby, *Biomed. Mass Spectrom.*, 1980, **7** (11), 597.

[4] *Applications of New Mass Spectrometry Techniques in Pesticide Chemistry*, ed. J. D. Rosen, John Wiley & Sons, New York, USA, 1987.

[5] G. A. Warburton and J. E. Zumberge, *Anal. Chem.*, 1983, **55**, 123.

[6] M. E. Rose and R. A. W. Johnstone, *Mass Spectrometry for Chemists and Biochemists*, Cambridge University Press, Cambridge, UK, 1982.

[7] J. R. Chapman, *Practical Organic Mass Spectrometry*, John Wiley & Sons, Chichester, UK, 2nd edn, 1994.

[8] W. H. McFadden, *Techniques of Combined Gas Chromatography/Mass Spectrometry: Applications in Organic Analysis*, Wiley-Interscience, New York, 1973.

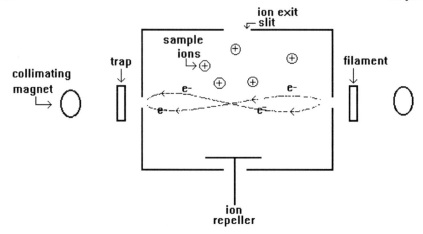

Figure 3.1 *Electron impact ionization source*

positively charged ion with one unpaired electron (a radical cation), the $M^{+\bullet}$ ion, from which the molecular weight of the sample can be inferred directly, according to Equation (3.1).

$$M \quad \overset{-e}{\rightarrow} \quad M^{+\bullet} \tag{3.1}$$

where M is a sample molecule,
e is an electron, and
$M^{+\bullet}$ is the molecular ion resulting from loss of one electron from M.

The $M^{+\bullet}$ ion can remain intact, or if there is sufficient energy remaining and the internal covalent bonds in the molecule are susceptible to cleavage, then fragmentations can occur. These fragment ions can in turn fragment if adequate energy exists, and so on. Most compounds tend to exhibit both the $M^{+\bullet}$ and fragment ions, and the relative ratio of the two depends on the energy required to break the molecule's internal bonds. For example, some C–C bonds tend to fragment readily and so most alkanes show only weak molecular ions accompanied by intense fragmentation. When a chemical bond cleaves to generate two or more fragments, then the charge is usually retained by the fragment with the lowest ionization potential, which is the route demanding the least energy.

There is a rule which states that radical (or odd-electron) ions, such as the molecular ions, fragment either by loss of a radical moiety or by loss of an even-electron unit, whereas even electron ions, such as some of the ions formed as a result of fragmentation of the molecular ions, fragment only by loss of even-electron units. Examples of radical groups include halide, alkyl, and hydroxyl radicals, and examples of even-electron units include H_2O (corresponding ions appear 18 mass units lower than their precursor ions in the spectrum), C_2H_4 (loss of 28 mass units), and CH_3OH (loss of 32 mass units). This principle is summarized in Equation (3.2).

Radical cations (odd-electron ions)

$$M^{+\bullet} \rightarrow F_1^+ + (radical)^\bullet$$

$$or \quad M^{+\bullet} \rightarrow F_1^{+\bullet} + even\text{-}electron\ unit$$

Cations (even-electron ions)

$$F_1^+ \rightarrow F_2^+ + even\text{-}electron\ unit \qquad (3.2)$$

where $M^{+\bullet}$ represents the molecular ions with an unpaired (odd) electron resulting from the loss of an electron,
F_1^+ represents an even-electron primary fragment ion,
$F_1^{+\bullet}$ represents an odd-electron primary fragment ion, and
F_2^+ represents an even-electron secondary fragment ion.

All of the sample-related ions created are accelerated out of the source, often with the help a positive (repeller) potential on the source. The source is heated to prevent any condensation of sample molecules, and the standard operating temperatures cover a wide range from 100 and 300 °C, depending on the type of sample under investigation.

The alternative process of electron capture to create a negatively charged radical anion [Equation (3.3)] is about 100 times less probable, and so EI ionization, in practice, is a positive ionization process.

$$M \quad \overset{+e}{\underset{\rightarrow}{}} \quad M^{-\bullet} \qquad (3.3)$$

where M is a sample molecule,
e is an electron, and
$M^{-\bullet}$ represents the molecular ions with an unpaired electron after gaining an electron.

Practical Operation of Electron Impact Ionization

Essential Requirements for Operation

A means of transporting the sample from the laboratory into the ionization source is necessary and in principle this could be a direct insertion probe (including those of the desorption ionization type) for solid samples, a gas chromatograph and interface for reasonably volatile samples in solution, a reservoir inlet system for non-polar, volatile samples and reference materials, or a particle beam (PB) liquid chromatography interface for more polar organic molecules in solution. The user needs to decide which inlet is both appropriate and available. In all cases, however, it is conceivable that at some stage a calibration of the m/z scale will need to be performed, and this is generally carried out using the reference inlet to admit a volatile reference compound continuously into the source.

A list of basic requirements for EI ionization is as follows:

- a functional mass spectrometer with an electron impact source fitted under a good vacuum, and generally set up according to the manufacturer's instructions;
- a reference or reservoir inlet, a suitable reference compound for calibration, *e.g.* heptacosafluorotributylamine [heptacosa, FC43 $(C_4F_9)_3N$], or perfluorokerosene [PFK, $CF_3(CF_2)_nCF_3$], and a syringe dedicated to loading the reference material;
- a direct insertion or desorption probe, a GC interface, or a particle beam interface;
- if a direct insertion probe is being used, a supply of disposable, capillary sample vials may be necessary (check the manufacturer's instructions) as well as a sample block to hold the vials (some blocks can be heated to aid evaporation of the solvent), suitable solvents to dissolve the sample, and a syringe to load the sample into the vial;
- if the desorption ionization probe is being used, solvents and a syringe will be necessary to dissolve and load the sample;
- if GC or particle beam LC interfaces are being used, then the necessary gases and solvents will be required; these topics are dealt with in more detail in Section 4 of this Chapter;
- chemical standards which are known to ionize well and produce reliable, reproducible results so that the system can be optimized and validated before any samples are analysed; some suggestions are made at appropriate places in the following text.

Setting up Electron Impact Ionization–Mass Spectrometry

The Filament and the Trap. For most mass spectrometers, the EI source will be **trap regulated**, rather than emission regulated. This means that the current through the **filament** (from which the electrons are emitted) is controlled by monitoring the electrons which succeed in reaching the trap situated at the opposite end of the ionization chamber to the filament, rather than by monitoring the total number of electrons emitted from the filament. A trap current of 200 μA is typical for a clean source, but when the source becomes contaminated, fewer electrons will be monitored at the trap and then the trap current may have to be set at a higher value to maintain sensitivity.

The filament is the most likely cause of any faults that may arise during EI operation. The filament should only be switched on when the operating vacuum has been reached, and indeed most mass spectrometers will not allow the user to do otherwise.

Problems which may occur include the following.

1. If the filament current is registering zero, then either the filament has burnt out or there is a loose connection to the filament.
2. If the filament is in contact with the ionization chamber then the electron energy will be shorted and the trap current will be zero.

If these or any other problems arise immediately after cleaning the ionization source, then the alignment of the filament should be rechecked, or the filament replaced as sometimes a filament can distort and bend slightly when the source is heated. Care should be taken also to ensure that the source magnets are correctly reinstated with the opposite poles facing.

The EI source should be cleaned whenever necessary, which in the extreme case is when the performance drops off to the extent that no useful data can be obtained. Great care should be taken when cleaning and handling the source; only the appropriate tools and cleaning materials should be used, and care must be taken not to touch the individual parts of the source by hand, by wearing nylon gloves or using tweezers. When reassembling the EI source, it is of vital importance to ensure that the filament is aligned correctly because if it is not, none of the electrons will reach the trap. In such a case the trap current might be raised and although more electrons would be emitted from the filament, still none would reach the trap and eventually the filament would burn out. A usable filament current would be 3–4.5 A, but will vary from one source to another.

Tuning and Optimizing the EI Source. The EI ionization **source temperature** is usually set in the region 180–220 °C for general purposes. The EI source should be tuned, optimized and calibrated before any samples are analysed, and the easiest way of doing this is to continuously introduce a liquid reference material into the source from a reference or reservoir inlet. Reservoir inlets are of many different shapes and forms; their internal capacities can vary from a few microlitres to over a litre, and some have the option of being heated. References inlets are usually dedicated to the analysis of reference compounds to avoid cross-contamination with samples, and are generally of a few microlitres in capacity. The reference inlet should be filled following the instructions supplied with a reference compound such as heptacosa, or alternatively PFK if the reference inlet is of the type that can be heated. Several microlitres of the neat liquid can usually be introduced into the reference inlet by means of a syringe, and it is a good idea to set aside one syringe specifically for this purpose, again to avoid cross-contamination of samples and reference materials. The flow from the reference inlet into the ionization source is controlled carefully by a restriction so that the vacuum gauge and the rather delicate filament of the ionization source are not subjected to vast fluctuations in pressures and also so that the reference material is consumed slowly and steadily.

With the reference material – now in its gaseous state – in the ionization source, a number of ions, say three or four, spread throughout the *m/z* range of interest, should be selected and optimized. Suitable heptacosa ions would be those at *m/z* 69, 219, 502, and 614, while appropriate PFK ions could include *m/z* 69, 131, 331, and 605. When optimizing these ions, the peak shape and resolution as well as the intensity should be considered. In EI it is usual to work with at least unit resolution. The ratio of the intensities of these ions should remain constant from day to day on the same mass spectrometer, and should be in accordance with the manufacturer's recommendations. If these ions vary substantially then remedial action should be taken to restore the operating conditions and the most common answer is to clean the source.

Once tuned up, and with the reference gas still in the source, a data acquisition can be made and the *m/z* range calibrated. Typical spectra for both heptacosa and PFK are illustrated in Figures 3.2 and 3.3 respectively, and usually a calibration up to *ca. m/z* 600 for heptacosa or *ca. m/z* 1000 for PFK is possible. These ranges are usually adequate for the types of sample that are analysed by EI. However, if certain samples demand a higher *m/z* calibration, then an alternative calibrant should be used, and admitted from the direct insertion probe if volatility becomes an issue. Several of the fluorinated triazines provide usable ions as high as *ca. m/z* 1500, although any suitable material which produces ions that are evenly spaced throughout the *m/z* range of interest, and for which the *m/z* values are known accurately, can be employed successfully. When calibrating in EI mode, the calibration is made to extend to *m/z* 40, which is sufficiently low to monitor the majority of fragment ions and also to allow the sample spectra to be compared with spectral libraries, because many of the entries in the commercially available libraries have *m/z* 40 for the lower limit of their spectra. Occasionally even lower *m/z* values need to be acquired, depending on the samples under scrutiny and their fragmentation patterns.

With the source tuning optimized and a good calibration in hand, the system is ready for samples to be analysed. If sensitivity is an important criterion, then it is useful at this stage to run a standard sample as a sensitivity check. This standard should be analysed by the same method of sample introduction that will be used for the samples, *e.g.* through the GC or by the probe, and can be any compound whose response is known by the operator.

Accurate Mass Measurements. Heptacosa and PFK both contain numerous fluorine atoms, which is a monoisotopic, mass deficient atom. This makes the calibration *m/z* values slightly mass deficient which can prove useful if **accurate mass measurements** are being performed. In such cases both the sample and the reference material need to be in the ionization source simultaneously, and as most organic compounds are mass sufficient, the reference ions do not interfere with the recording of the sample ions and so good measurements can be made.

Electron impact ionization is traditionally the method of choice for performing accurate mass measurements, although there is no reason why

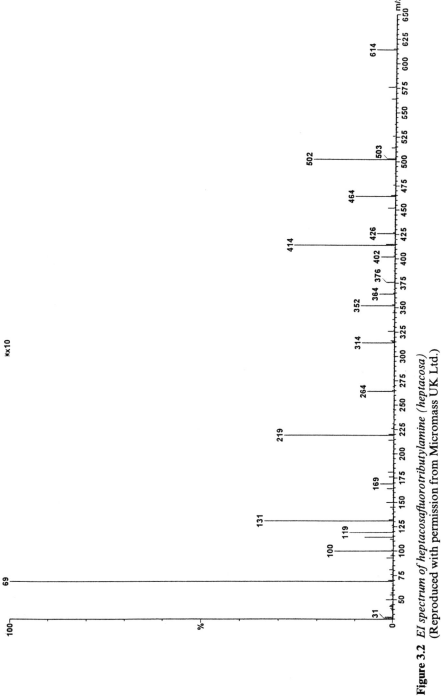

Figure 3.2 *EI spectrum of heptacosafluorotributylamine (heptacosa)* (Reproduced with permission from Micromass UK Ltd.)

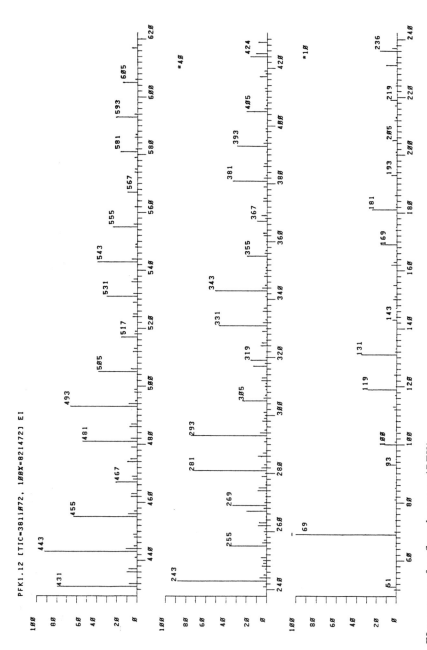

Figure 3.3 *EI spectrum of perfluorokerosene (PFK)*
(Reproduced with permission from Kratos Analytical)

other ionization methods cannot be used, as is often the case. EI generates molecular ions with average signal intensities, and so there is little chance of these ions being overloaded and hence erroneous readings being made. It is also quite straightforward under EI conditions to be able to admit the reference compound, through the reference inlet, simultaneously with the sample, from either the probe or the GC interface. In this way the sample analysis is not interfered with, and the relative intensities of the reference ions and sample ions can be adjusted simply, quickly and independently to obtain ion signals of approximately equal intensities, which is the ideal situation for accurate mass measurements. With practice, measurements should be achievable to 5 ppm or better.

When performing accurate mass measurements, the resolution of the mass spectrometer is raised wherever possible, and so magnetic sector mass spectrometers are commonly used. The resolution does not affect the accuracy of the results, but if more than one isobaric component is present, and each has a different molecular formula, then the resolution will separate the components and enable individual mass measurements to be made.

High resolution mass spectrometry is often required in the area of dioxin analysis, where it is successfully used in conjunction with GC–MS.[1,2,3]

The Analysis of Samples Using Direct Introduction Methods

The Reservoir Inlet System

If the samples to be analysed do not require any prior chromatographic separation and are quite volatile, then a suitable reservoir inlet can be used to admit such samples into the EI ionization source, in very much the same way that the reference materials were introduced. The type of samples that are usually analysed in this way are of the hydrocarbon type and are of high volatility; thus this technique is popular with the oil and petroleum industries.[3] Such samples are quite likely to contain mixtures of many similar and isomeric compounds, and at times some degree of fractional distillation may occur if the inlet is heated controllably.

The Direct Insertion Probe

The direct insertion probe is a very common method of introducing samples into the EI source. Usually a solid sample is dissolved in an appropriate solvent (say, 1 ng to 1 µg µL^{-1}) and 1 or 2 µL taken up by syringe and deposited into a glass or quartz capillary tube of approximately 2 cm length. The capillaries are sealed at one end and the glass ones can either be purchased commercially or made quite easily by rotating a portion of a melting point tube in a moderately hot flame. The capillaries are disposable as they are really too small to be cleaned out properly, and so an adequate supply should be accumulated and maintained. Once the sample and solution are in the capillary, the capillary should be set aside until the vast majority of the solvent

sample vial ———————→

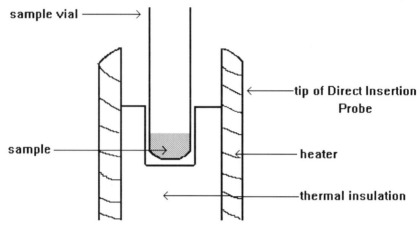

tip of Direct Insertion
Probe

sample ——————————

heater

thermal insulation

Figure 3.4 *Typical direct insertion probe*

has evaporated. This can be achieved at room temperature if the solvent is a volatile one, or by heating gently if the solvent is not so volatile or time is at a premium. It is possible to purchase or have made small racks where a number of these glass capillary vials can be placed and several samples prepared at once. Some of these racks are heated, or can be placed on a hot surface depending on the material from which they are constructed. If the sample-containing vial is heated too quickly, or inserted into the ionization source before most of the solvent has been allowed to evaporate, then the liquid within the vial will 'bump' out, leaving the vial empty.

Solid samples can also be applied to the bottom of a vial by use of a thin wire, but care should be taken that only the smallest amount of sample is transferred, otherwise the ionization source will soon become contaminated and the spectra produced will be saturated, leading to possible distortion of the isotope patterns. Liquid samples that are not sufficiently volatile for the reservoir inlet can also be inserted directly into the bottom of the capillary.

Once the sample is prepared, the vial can be loaded into the end of the probe. Most probes (see Figure 3.4) have a heater so that the involatile samples can be heated once inside the ionization source to aid vaporization, and so there should be good thermal contact between the sample vial and the heater unit of the probe. Quartz wool or tungsten foil have been used successfully for this purpose. The probe will probably have a cooling device which may consist of an inlet and an outlet for water to be passed through the probe. This allows controlled heating of the probe, and helps cool down the probe in between successive sample analyses. Before inserting the probe into the ionization source, the water supply (if present) should be connected and the probe reasonably cool. The probe insertion procedure can then be followed, which will normally consist of putting the probe into the direct insertion probe lock, pumping out the space around the probe, and then opening the lock to allow the probe to enter the source. In this way the source vacuum is maintained throughout the operation. Often the filament is switched off during this procedure, in case the source vacuum is unfavourably disturbed.

With the probe in position, the heat from the source and the filament may well be sufficient to vaporize the sample, and if so, sample-related ions will be detected immediately. Unless a sample is being admitted for tuning purposes, it is advisable to start a data acquisition as soon as the probe is in position, before any heat is applied, so that all of the components in the sample are detected and recorded. With the data acquisition under way, heat can be applied to the probe until the upper limit of the probe's temperature has been reached, if necessary. The direct insertion probe will have an upper temperature limit which will be defined in the manual supplied (possibly *ca.* 500 °C), and care should be taken not to try to exceed this limit. Direct insertion probes can be heated by using a temperature ramp, *e.g.* 20 or 50 °C min^{-1}, or the temperature can be controlled manually and incremented as and when the user thinks necessary, depending on the content of the spectra being generated.

Spectral Interpretation

An EI spectrum can contain vast amounts of information and so it is necessary to devote some considerable time to its interpretation.[9,10,11]

First the **molecular ion** must be designated, if possible. To do this, the higher *m/z* end of the spectrum should be scrutinized and any possible candidates considered. Often the theoretical molecular weight of the sample is known and it is a matter of confirming whether or not the appropriate ion is present. Not only the *m/z* value of the molecular ions should be noted but also whether or not it is an odd or an even number. An odd molecular weight indicates an odd number of nitrogen atoms in an organic (C, H, N, and optionally O containing) compound (*i.e.* 1, 3, 5, *etc.*) and an even molecular weight implies an even number of nitrogen atoms (*i.e.* 0, 2, 4, 6, *etc.*); see Chapter 1, Section 3.

Next on the checklist is the **isotope pattern** of the molecular ion region. A standard organic sample containing carbon and hydrogen atoms will have a dominant ^{12}C isotope, accompanied by a number of ^{13}C isotopes at intervals of one mass unit higher up the *m/z* scale. The ^{13}C isotopes will be of decreasing intensity, with the first one being (1.11 × *n*) % of the ^{12}C molecular ions, where *n* is the number of carbon atoms present in the compound (Chapter 1,

[9] H. Budzikiewicz, C. Djerassi and D. H. Williams, *Structure Elucidation of Organic Compounds*, Vols. I and II, Holen Day, San Francisco, USA, 1964.
[10] D. H. Williams and I. Fleming, *Spectroscopic Methods in Organic Chemistry*, McGraw Hill, Berkshire, UK, 2nd edn, 1973.
[11] F. W. McLafferty and F. Turecek, *Interpretation of Mass Spectra*, University Science Books, CA, USA, 4th edn, 1993.

Section 2). This is probably the most common isotope pattern observed in most laboratories. However, if chlorine and bromine atoms are present, then the isotope distribution will be quite distinctive and very informative (see Chapter 1, Figure 1.6). Sulfur compounds too have a distinctive if rather more discrete isotope pattern.

When a candidate for a molecular ion has been identified, it may well be possible to detect some **fragment ions**. Mass differences between the molecular ion and other ions, and also between all the other ions should be calculated and any of these mass differences that relate to the loss of common functional groups may well provide clues to the structure of an unknown compound, or substantiate the theoretical structure expected. Common mass losses include 15 (loss of a methyl group), 18 (loss of water), 28 (loss of CO), 32 (loss of methanol), and 44 (loss of CO_2), to name just a few (see Chapter 1, Figure 1.12 for further information). Some ions in the spectrum will indicate the structural nucleus of the compound under investigation: for example, an ion at m/z 77 is characteristic in most cases of $C_6H_5^+$ ions which indicate an aromatic compound; a series of ions of m/z 29, 43, 57, 71, 85, *etc.* affirm the presence of a hydrocarbon. Some methyl esters such as methyl stearate undergo a rearrangement to generate very diagnostic ions at m/z 74 ($C_3H_6O_2$). Entire books have been dedicated to the fragmentation pathways of various compounds[9,11] and should be consulted if necessary as it is beyond the scope of this monograph to detail the fragmentation behaviour of every class of organic compound.

Two examples of EI spectra are presented to illustrate the principles outlined above, albeit with quite uncomplicated molecules, *viz.* 2,6-dimethyl-phenol and 2,6-dimethylaniline, depicted in Figures 3.5 and 3.6 respectively. 2,6-Dimethylphenol has the molecular formula $C_8H_{10}O$, and a monoisotopic molecular weight of 122.0729. From the spectrum in Figure 3.5, the molecular ions $C_8H_{10}O^{+\bullet}$ at m/z 122 can be seen, accompanied with the much less intense ^{13}C isotope ions at m/z 123. A small signal is apparent at m/z 121, indicating loss of a labile proton from the molecule. The next significant ions appear at m/z 107 and correspond to the loss of a methyl (-15) group from the molecular ions. The ions at m/z 91 and 77 can be attributed to and are characteristic of the benzyl and aryl aromatic ions corresponding to $C_7H_7^+$ and $C_6H_5^+$, respectively; m/z 91 is usually responsible for fragment ions at m/z 65 ($C_5H_5^+$); m/z 78 and 79 ($C_6H_6^{+\bullet}$ and $C_6H_7^+$) also confirm the aromatic nature of the compound.

2,6-Dimethylaniline has a molecular formula of $C_8H_{11}N$ and a monoisotopic molecular weight of 121.0888. The molecular ions $C_8H_{11}N^{+\bullet}$, can be seen at m/z 121, and again there is a loss of a labile proton to generate ions at m/z 120 (Figure 3.6). As 121 is an odd number, this confirms the fact that there are an odd number of nitrogen atoms in the molecule. The isotope pattern of the molecular related ions is consistent with a standard organic compound. Fragment ions at m/z 106, consistent with loss of a methyl group, can be seen, and the ions at m/z 91, 79, 78, 77, and 65 confirm the aromatic nature of this compound, as with the previous compound.

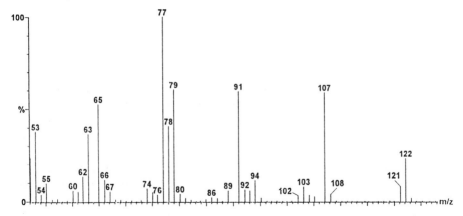

Figure 3.5 *EI spectrum of 2,6-dimethylphenol (MW 122 da)*
(Reproduced with permission from Micromass UK Ltd.)

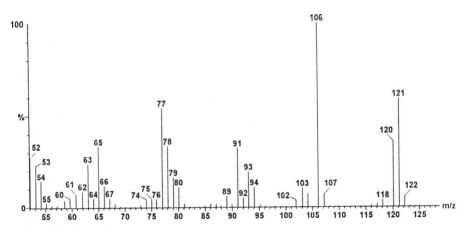

Figure 3.6 *EI spectrum of 2,6-dimethylaniline (MW 121 da)*
(Reproduced with permission from Micromass UK Ltd.)

There are several commercially available **libraries** containing many thousands of spectra which have been built up over the years, and most modern data systems have the means of accessing, searching, and comparing these spectra with the acquired spectra in seconds. This is an excellent way of interpreting and confirming EI spectra, and there are possibilities to extend this to searching sub-libraries, for example by limiting the library to those spectra originating from hydrocarbons alone, or a sub-library whose entries include those compounds containing *inter alia* a single chlorine atom. The standard libraries available can also be updated if a new compound is identified. The library search routines have the advantage of automation; a search method could be included in a general processing routine which is carried out at the conclusion of a GC–MS run.

3 Chemical Ionization

The Principles of Chemical Ionization

The technique of CI[12,13,14] generally uses the same, or at least a similar source to that employed by EI ionization. The filament and trap are still present, but now a gas is introduced into the ionization chamber, increasing the source pressure by a factor of approximately ten. The electron beam emitted from the filament now experiences greater difficulty in reaching the trap due to the higher source pressure and the increased chance of collisions with gas molecules and ions. In CI operation the filament current is therefore regulated by monitoring the filament output rather than the current monitored at the trap.

The **reagent ions** are formed from the gas molecules by collision with the electron beam emitted from the filament, and it is much more probable that these reagent ions, which are present in a vast excess to the sample molecules, react with the sample molecules to form product ions rather than the sample molecules being ionized directly by electron impact. To maintain a good gas pressure and increase the probability of the reagent ions reacting with the sample molecules, the CI source needs to be much more gas-tight than for EI operation. Some sources can be adapted for either EI or CI work by having ion exit slits of different sizes, with the smaller exit slit being used for CI analyses and the larger one for EI analyses. In some cases the slit can be manipulated while the source is under vacuum, or in other instances the source must be vented to change the slit aperture. Alternatively, sources exist which are combined EI/CI sources, and as such their performance may be slightly compromised; for example, the EI sensitivity may be a factor of *ca.* two less than would be expected from an EI only source, but their redeeming feature is the ease with which the two techniques can be interchanged.

CI is effective in both the **positive and negative ionization** modes, and in both cases is classed as a 'soft' ionization technique because the spectra, in contrast to EI spectra, are dominated by quasi-molecular ions, with very little fragmentation taking place. CI is often used in cases where EI fails to produce a molecular ion, perhaps due to excessive fragmentation, thus making it difficult to glean the molecular weight of the sample from the EI spectrum.

The reagent ions generated are very much dependent on the gas admitted into the source. The most commonly used gases are **methane, iso-butane** (2-methylpropane), and **ammonia**, and each has advantages and disadvantages for different application areas. If methane is the reagent gas, the major reagent ions formed by **positive ionization** electron impact are generally considered to be CH_5^+ and $C_2H_5^+$; *iso*-butane ionizes to generate $C_3H_7^+$ and $C_4H_9^+$ ions; ammonia forms monomeric and dimeric NH_4^+ reagent ions. The most

[12] M. S. B. Munson and F. H. Field, *J. Am. Chem. Soc.*, 1966, **88**, 2621.
[13] K. R. Jennings, in Chapter 12, *Gas Phase Ion Chemistry*, ed. M. T. Bowers, Academic Press, New York, USA, 1979.
[14] H. Budzikiewicz, *Mass Spectrom. Rev.*, 1986, **5**, 345.

common reaction pathway for these reagent ions is to protonate the sample molecules and produce dominant MH^+ ions leaving a neutral species such as CH_4 or C_2H_4, and this will happen as long as the proton affinity of the sample is greater than that of the reagent gas. If this is not the case then attachment, or adducting, can occur, and $(M + C_2H_5)^+$, $(M + C_4H_9)^+$ and $(M + NH_4)^+$ ions may be formed and detected. For example, the proton affinity of most oxygen-containing organic compounds (*e.g.* aldehydes, ketones, esters) is greater than the proton affinity of the $C_2H_5^+$ and $C_4H_9^+$ reagent ions, but less than the proton affinity of the NH_4^+ reagent ions. Most oxygen-containing organic compounds therefore will form MH^+ ions when methane or *iso*-butane is used as the reagent gas, but when ammonia gas is present, these compounds will generate the $(M + NH_4)^+$ species. In fact, only the most basic of organic compounds such as amines and some nitrogen heterocycles will be protonated by ammonia reagent gas. At the other extreme, for those samples with very low proton affinities, such as aliphatic hydrocarbons of the general formula C_nH_{2n+2}, it is possible for the reagent gas ions to abstract a proton to generate $(M - H)^+$ sample-related ions.

Although **negative ion**[6,7] production under EI conditions is not favourable because the electron energy tends to be too high to permit electron capture reactions to take place, the scenario is different when operating under CI conditions. The presence of a gas at relatively high pressures provides a buffered environment in which the electrons are considerably decelerated, and those molecules which have functional groups appropriate for electron capture (*i.e.* electronegative atoms or functionalities) can do so to form **negatively charged radical molecular ions**, $M^{-\bullet}$. In such cases of **electron capture**, the gas is not strictly a reagent gas because it is not reacting with the sample.

If electron capture is not possible, then a reaction between the sample molecules and **negatively charged reagent gas ions** can be employed, although such experiments are less commonly encountered.[14] For example, some halogenated compounds can be admitted as reagent gases into the ionization source where they will undergo ionization to produce F^-, Cl^-, Br^-, or I^- ions. Chloroform and dichloromethane, for instance, will generate Cl^- ions. It is possible for these halide ions to react with certain sample molecules and produce attachment or adduct ions, *e.g.* $(M + Cl)^-$. Some reagent ions, *e.g.* those containing carboxylate (CO_2^-) or hydroxylate (OH^-) groups, will generate deprotonated molecular anions, $(M - H)^-$, depending on whether the relative proton affinities of the reagent ions is greater than that of the sample molecules.[6,7,15] Negative chemical ionization can lead to a variety of results and it is important to devise an experiment that will lead to the best results for the particular compounds being analysed, and to interpret the spectra carefully.

A summary of the more frequently encountered chemical ionization product ions is presented in Table 3.1.

[15] A. P. Bruins, in *Advances in Mass Spectrometry*, ed. J. F. J. Todd, John Wiley & Sons, Chichester, UK, p. 119, 1985.

Table 3.1 *Frequently encountered chemical ionization product ions*

	Positive chemical ionization	
Reagent gas	*Major reagent ions*	*Possible product ions*
methane, CH_4	CH_5^+, $C_2H_5^+$	MH^+, $(M-H)^+$, $(M+C_2H_5)^+$
iso-butane, C_4H_{10}	$C_3H_7^+$, $C_4H_9^+$	MH^+, $(M-H)^+$, $(M+C_4H_9)^+$
ammonia, NH_3	NH_4^+, $(NH_3)_2H^+$	MH^+, MNH_4^+
	Negative chemical ionization	
Ionization mechanism	*Possible product ions*	
Electron capture	$M^{-\bullet}$	
Deprotonation	$(M-H)^-$	
Adduct formation	$M(adduct)^-$	

The relative sensitivities of EI (EI+), positive CI (CI+), and negative CI (CI−) are all very much dependent on the type of sample being analysed. If sensitivity is an issue, then all of these techniques should be experimented with to find the most sensitive and appropriate method for *that particular compound*. Often, both EI and CI are carried out on the same sample, as the two ionization methods produce complementary information and together have a valuable input into the molecular weight and structural determination of a sample. Some mass spectrometers allow alternate scans to be acquired in EI followed by CI (positive and negative modes) throughout the entire analysis, which for a sample introduced via a gas chromatograph provides two or three readily comparable chromatograms from a single injection.

Practical Operation of Chemical Ionization

Essential Requirements for Operation

A means of transporting the sample from the laboratory into the ionization source is necessary, and basically the options are the same as detailed for EI operation (Chapter 3, Section 2) *i.e.* a direct insertion probe (including those of the desorption ionization type) for solid samples, a gas chromatograph and interface for reasonably volatile samples in solution, a reference inlet for calibration materials, or a particle beam (PB) liquid chromatography interface for more polar organic molecules. For positive CI, the instrument is usually calibrated in the EI mode of operation, and then switched over to CI, because very few reference materials will produce a spectrum with a good number of ions spread evenly throughout the *m/z* range. For negative CI, the *m/z* scale can be calibrated with negative ions generated from either heptacosa or perfluorokerosene introduced via the reference inlet, in the same manner described under EI operation (see Chapter 3, Section 2).

A basic list of requirements for CI ionization is as follows:

- a functional mass spectrometer with a clean chemical ionization source fitted operating under a good vacuum, and generally set up according to the manufacturer's instructions;
- a supply of reagent gas, or gases, the most common ones being methane, *iso*-butane (2-methylpropane), and ammonia;
- a reference inlet, a suitable reference compound for calibration, *e.g.* heptacosafluorotributylamine [heptacosa $(C_4F_9)_3N$] or perfluorokerosene [PFK, $CF_3(CF_2)_nCF_3$], and a specially designated syringe to load the reference material;
- a direct insertion or desorption probe, or alternatively a GC interface or a particle beam interface (NB. some reservoir probes are not suitable for CI work);
- if a direct insertion probe is being used, a supply of disposable, capillary sample vials may be necessary (check the manufacturer's instructions) as well as a sample block to hold the vials (some blocks can be heated to aid evaporation of the solvent), suitable solvents to dissolve the sample, and a syringe to load the sample into the vial;
- if a desorption chemical ionization probe is being used, a range of solvents for dissolving samples and a syringe for loading the sample onto the probe;
- if GC or particle beam LC are being interfaced, then the necessary gases and solvents will be required: these topics are dealt with in more detail in Section 4 of this Chapter;
- chemical standards which are known to ionize well and produce reliable, reproducible results so that the system can be optimized and validated before any samples are analysed, *e.g.* amyl acetate, benzophenone, and xylene for CI+ and octafluoronaphthalene for CI− work.

Setting up Chemical Ionization–Mass Spectrometry

First of all, a reagent gas should be selected. Methane, *iso*-butane and ammonia are the three most widely used gases for positive CI analyses, and the important difference between them is their proton affinities:

methane < *iso*-butane < ammonia

Methane and *iso*-butane tend to be used with non-polar organic compounds of low proton affinities (such as hydrocarbons), which will usually protonate to produce MH^+ ions or deprotonate to generate $(M-H)^+$ ions. Ammonia is useful for more polar organic compounds which, in the case of strongly basic amines, will produce MH^+ ions, or in the case of oxygen-containing molecules may generate preferentially the MNH_4^+ ions.

More acidic compounds such as carboxylic acids may produce superior results under negative ionization conditions, yielding RCO_2^- ions, as do compounds containing electronegative atoms such as Br, Cl, and F. The choice of reagent gas is not so critical for electron capture negative ionization work because in this case, a gas is required in the source to slow down the fast moving electrons in order to increase the probability of electron capture reactions.

Often samples will contain a mixture of widely differing molecules and in these cases a gas must be chosen which may produce excellent data for some of the components but only mediocre spectra for the remainder, unless the operator has sufficient sample to repeat the analysis with a second reagent gas. When operating under CI conditions, it should be remembered that in the continual presence of a reagent gas, the source will become contaminated much more quickly than in EI operation, especially if methane or *iso*-butane is being used. Ammonia does not have such a detrimental effect on the source. There will be a continual presence in the source of the reagent gas ions, which will all appear in the mass spectrum if the scanning range is set at a sufficiently low *m/z* value; specifically in the case of *iso*-butane, where the *m/z* 57 ions may well interfere with sample-related ions, and the low *m/z* limit of the data acquisition should be set to *m/z* 60 to avoid generating unnecessarily large data files.

So, with the reagent gas selected the gas cylinder should be connected to the appropriate port on the mass spectrometer. Before admitting the gas into the source, the gas lines should be pumped out to remove any air or residual traces of other gases which may be present. Once set up, it is preferable to leave the gas cylinder in position for future work, and some mass spectrometers have the facility for having several gas cylinders connected simultaneously so that the operator does not have the tiresome task of changing cylinders and connecting the gas lines for each different type of CI analysis.

The next step is to seal the source as much as possible to help ensure that it is gas-tight, and so the direct insertion probe should be put in place (taking care with the insertion lock) to block off this inlet. The probe should always be in position in the source during CI operation, even when the samples are being introduced through the GC. There may be other source re-entrants that should have been sealed off before fitting the source in the source housing, and it is advisable to check before commencing any CI work.

Now, with the gas lines pumped out and the gas cylinder opened, the reagent gas can be admitted into the ionization source until the source pressure reading has increased by a factor of approximately ten. The source tuning parameters should be optimized initially on the reagent gas ions, and then with a known sample such as benzophenone or xylene for CI+ work, or octafluoro-naphthalene for CI− studies. Parameters to check when tuning include the gas pressure, the emission of electrons from the filament, the energy of these electrons, and the repeller which assists the sample ions to exit from the source. For CI analyses, the source temperature is usually set a little lower than for EI work in order to minimize fragmentation and optimize the quasi-molecular ions. Typical CI operating temperatures are in the region of 150 to 180 °C rather than the EI values of 180 to 220 °C, although this does depend on the samples under scrutiny.

If methane reagent gas is being used for positive CI analyses, the reagent ions at m/z 17 (CH_5^+) and 29 ($C_2H_5^+$) should be present in approximately equal proportions. With these ions displayed on an oscilloscope or a real-time data system display, the gas pressure should be adjusted for maximum signal strength. A hydrocarbon sample is most appropriate for tuning purposes under these conditions, and xylene can be introduced into the ionization source through reference inlet for this purpose. The MH^+ ions at m/z 107 should be optimized at the expense of any $M^{+\bullet}$ ions at m/z 106 using the source parameters.

With *iso*-butane reagent gas, positive CI conditions are generally optimized when the $C_3H_7^+$ ions at m/z 43 are approximately half the intensity of the $C_4H_9^+$ ions at m/z 57. When this has been achieved, further tuning with hydrocarbon-type standards can be undertaken.

Ammonia as positive CI reagent gas generates ions at both m/z 18 and 35, corresponding to monomeric and dimeric (NH_4^+) and [$(NH_3)_2H^+$)] ions respectively. Optimum CI data are usually obtained after maximizing the ions at m/z 35. An oxygen-containing organic compound such as the ester amyl acetate (MW 130) or the ketone benzophenone (MW 182) can be used as a tuning standard, and these samples should produce both MH^+ and MNH_4^+ ions.

For **negative CI** electron capture operation, ammonia is often the preferred reagent gas, although methane and *iso*-butane can also be used, as ammonia contaminates the source less quickly. The reagent gas ions should be optimized as described above under CI+ conditions, and then the mass spectrometer switched over to CI− operation. The optimum level of reagent gas and all other tuning parameters should be checked in negative ionization mode using a suitable standard compound such as heptacosa. Again the source pressure will probably be a factor of *ca.* 10 higher than under EI conditions, when the gas has been admitted and optimized. When tuning with heptacosa, the ions at m/z 452 and 633 should both be of significant intensity. Heptacosa is also a good standard for calibration purposes when operating in CI−.

Once the CI ionization source has been tuned adequately in either positive or negative ionization mode, it is advisable to evaluate the sensitivity of the system with a known compound which should be analysed at a known concentration. Results from this compound can be compared on a day-to-day basis and any major variations investigated immediately and rectified. The standard can be admitted into the source using either the direct insertion probe or a GC interface, although it is advisable to use the same method of introduction that any subsequent analyses will require, to ensure that the complete system is checked out.

For these purposes, benzophenone is suitable for CI+ work and octafluoronaphthalene for CI− studies. Figure 3.7 shows the ammonia CI+ spectrum of benzophenone (MW 182), with MH^+ ions at m/z 183 and MNH_4^+ ions at m/z 200, and Figure 3.8 illustrates electron capture under CI− conditions for octafluoronaphthalene, with the $M^{-\bullet}$ ions at m/z 272 dominating the spectrum. The ultimate performance of the mass spectrometer will vary very much from one instrument to another, but it is important that the instrument being used is operating to the best of its potential, and so it is necessary for the

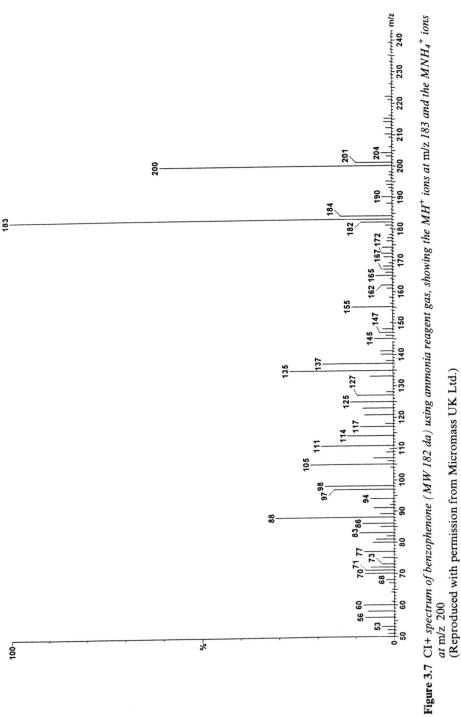

Figure 3.7 CI+ spectrum of benzophenone (MW 182 da) using ammonia reagent gas, showing the MH⁺ ions at m/z 183 and the MNH₄⁺ ions at m/z 200
(Reproduced with permission from Micromass UK Ltd.)

Figure 3.8 *CI — spectrum of octafluoronaphthalene (MW 272 da) showing M⁻• ions at m/z 272*
(Reproduced with permission from Micromass UK Ltd.)

operator to determine exactly what this potential is, and to achieve this target before commencing any sample analyses.

The Analysis of Samples Using Direct Introduction Methods

The Direct Insertion Probe

To load the samples into vials and use the direct insertion probe, the procedure described in detail under electron impact ionization (Chapter 3, Section 2) should be followed.

The Desorption Chemical Ionization Probe

The desorption chemical ionization (DCI) probe is an alternative to the direct insertion probe, and despite its title can be employed equally well in positive EI as well as both positive and negative CI modes of operation. The DCI probe is designed to cope with samples that are susceptible to thermal decomposition, although there is no reason why it cannot be used as a general means of introducing samples into the ionization source, *i.e.* as an alternative to the direct insertion probe.

The DCI probe tip is composed of a small platinum coil attached between two posts (see Figure 3.9), and as this arrangement is quite delicate, great care must be taken to ensure that the probe tip is not damaged in any way, especially when inserting and removing the probe through the insertion lock. Indeed, some DCI probes have a retractable probe tip which is used to protect the tip when transporting the probe through the vacuum lock, thus minimizing the chance of damaging the tip during this operation.

To use the DCI probe, 1–2 μL of the sample, in solution, are applied to the probe tip and after the solvent has been allowed to evaporate at room temperature, the probe is inserted into the source. Once in position, an acquisition is started and then an electrical current is passed through the platinum coil to heat the sample. This is a very efficient method of heating the sample directly, and DCI probes have the capability of very fast temperature ramping, the idea being that the sample is volatilized before it has chance to thermally decompose.

The types of sample which benefit most from a good DCI probe are the higher molecular weight, less volatile compounds including some of the natural product classes, (*e.g.* alkaloids, sugars, steroids), organometallics, peptides, and any very thermally sensitive compounds.

Before a sample is loaded onto the platinum coil, the coil should be cleaned by holding it in a naked flame or immersing it in a suitable solvent. Sample loading can be made either by using a syringe to place *ca.* 2 μL of the sample solution onto the coil, or by dipping the coil directly into the sample solution. This latter method is not highly recommended as some sample may end up on the support rods holding the platinum coil, and this may result in cross-contamination of samples. The strength of the sample solution depends on the

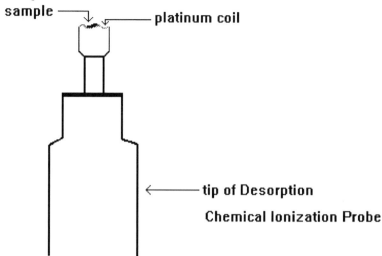

Figure 3.9 *Typical desorption chemical ionization probe*

amount of sample available, the sample's ability to be ionized and the performance of the mass spectrometer being used. For an initial experiment, a concentration of between 0.1 and 1 mg mL^{-1} should be tried. As volatile a solvent as possible should be used to dissolve the sample so that the time required to wait for the solvent to evaporate and leave a thin layer of sample covering the coil is minimized.

The rate at which the heater is ramped also varies from sample to sample. Some samples need to be heated as quickly as possible, and temperature ramps from 20 to 700 °C over several seconds are sometimes used. Other samples may contain mixtures from which components can be fractionally desorbed with slower temperature gradients. Usually the temperature ramp is achieved under data system control and so can be reproduced accurately.

4 Separation Methods Coupled to Electron Impact and Chemical Ionization

Gas Chromatography

GC–MS Interfacing

On-line chromatography–mass spectrometry is invaluable for the analysis of complex mixtures and **gas chromatography (GC)** was the first chromatographic method to be used successfully in combination with mass spectrometric studies. When GC was first coupled to mass spectrometers, electron impact and chemical ionization were the only ionization methods available, and the early GC–MS interfaces had to overcome the challenge of coupling packed GC columns, with a high flow of carrier gas, to an ionization source that could be operated only under a high vacuum. These days life is simpler as most GC analyses utilize the high resolution capillary columns which require only a modest carrier gas flow that can flow directly into the ionization source without breaking the vacuum. This treatise is limited to the more common, modern GC–MS interfaces of the latter type, rather than being an historical overview of the development of GC and GC–MS. For the interested reader, the historical aspect is a subject worth further investigation if time permits.[8]

Very briefly, GC provides a method of separating components in a mixture by eluting the sample with a carrier gas such as helium over a stationary phase which is contained within a column. To aid the elution of components with differing volatilities, an increasing temperature program is used, and to achieve this the GC column is held within an oven. If successful, a number of chromatographic peaks, each one several seconds wide, should emerge from the exit of the column and be detected. Conventional detectors include flame ionization detectors, electron capture detectors, and mass spectrometers. If the column terminates within the ionization source of a mass spectrometer, then mass spectra will be produced for all of the separated components. GC can be used equally well in conjunction with both electron impact and chemical ionization.

Mass spectrometers now arrive equipped with a GC interface and a gas chromatograph too, if this is available and has been requested on that particular model. Often the operation of the gas chromatograph is controlled through the data system of the mass spectrometer so that the GC and the MS are synchronized. The GC–MS interface is usually kept to as short a length as possible, which means that the GC is situated as close as possible to the mass spectrometer, with the interface reaching from the GC on one side to the ionization source on the other. The GC capillary column emerges from the GC oven directly into the GC interface and terminates within the ionization source, so the main task of the GC–MS interface is to heat a short length of the column evenly to a desired temperature in order to prevent any eluates condensing *en route* to the ionization source.

GC Columns

In practice the GC oven is equipped with a **column** suitable for the particular analysis to be undertaken, and the column is fitted to the **injection port** of choice. The suppliers of GC columns usually have extensive literature complete with examples of chromatography, so that when choosing a column there are often several options available. When selecting a GC column, the type and thickness of the stationary phase, and the column length and internal diameter must all be considered with respect to the sample type, the concentration of the components of interest and the complexity of the sample, in order to produce the optimum results. A short, very simplified list of some of the more commonly encountered stationary phases and their uses is given in Table 3.2.

The thickness of the **stationary phase** chosen depends on the volatility of the sample. Volatile compounds require quite thick films to retain them on the

Table 3.2 *Commonly encountered GC stationary phases*

Stationary phase	Sample type
100% dimethylpolysiloxane	general use, hydrocarbons, pesticides, steroids, derivatized sugars
5% phenyl, 95% methylpolysiloxane	hydrocarbons, alcohols, fatty acids, pesticides, steroids, polychlorobiphenyls
polyethylene glycol	essential oils, fatty acids, flavours, esters, amines, acids, glycols

column long enough to resolve them, while involatile compounds are better suited to thin film stationary phases, otherwise they will be retained too long on the column and the peak shape will suffer as a result. GC columns can 'bleed', or lose a little of the stationary phase, especially when heated up to their maximum temperature, and the thicker the stationary phase, the more column bleed one may expect. When coupled directly to a mass spectrometer, the stationary phase shed ends up in the ionization source and is ionized, just as all other eluents are. The ions associated with the stationary phase are usually siloxane-related and appear at *m/z* 207, 281, 355 *etc.*; they soon become a familiar sight to the mass spectroscopist! The intensity profile of these ions will increase directly with increasing GC oven temperature.

The **column length** is dependent on the complexity of the sample; for example, most samples can be separated effectively with a 25 m column, but very complex samples containing many closely eluting components will require a 50 m column.

The **internal diameter** of the column depends on the concentration of the sample to be analysed. Small diameter columns, *e.g.* 0.25 mm, offer high efficiency and are useful for the quantification of small amounts of sample, but on the other hand are easily overloaded. Columns of internal diameter 0.32 mm are recommended for on-column injections, whereby all of the solvent and sample are injected directly onto the column, while 0.53 mm columns are designed to replace the older style packed columns for high volume injections such as those used for purge-and-trap analyses. All of the columns require a continuous flow of gas through them, which constitutes the mobile phase. Helium is the gas of choice, being very nearly as efficient as hydrogen but much safer to handle. The smaller the internal diameter of the column, the lower the flow rate of gas required, and Table 3.3 provides a basic guide to the carrier gas flow rate suggested for different columns.

In most cases the choice of column does not affect the operation of the mass spectrometer.

GC Injectors

The **injector ports** available will vary from one GC to another, and usually there is more than one, and sometimes as many as three present on a gas

Table 3.3 *Suggested carrier gas flow rates for GC columns of varying internal diameters*

GC column internal diameter (mm i.d.)	Helium gas flow rate (mL min^{-1})
0.10	0.30
0.25	0.80
0.32	1.30
0.53	3.00

chromatograph. The manufacturer's literature should be referred to before using the GC as the guidelines given here are of necessity very general ones. The most frequently encountered injection types are the **split**, the **splitless**, and the **on-column** injection, and the choice depends on both the volatility and concentration of the sample. The split and splitless injector ports are usually maintained at a temperature of 250 °C, while the on-column injector is not heated, and indeed can sometimes be cooled. Table 3.4 provides a comparison and general guidelines to their operation.

High purity helium carrier should be used, and the whole system should be checked for gas leaks before use, paying particular attention to the injector and any associated nuts and septa. In order to check for leakage, the system must be sealed and pressurized, and then observed carefully to see if the pressure holds when the helium gas is temporarily switched off. It is important that air is not allowed to enter the column, as this will impair the stationary phase. GC columns should always have a suitable carrier gas flowing through before being heated.

GC Oven

The GC **oven** temperature is of importance. The temperature of the oven at the beginning of the run needs to be set appropriately depending on the injector port chosen and the boiling point of the solvent being injected, as does the temperature ramp and the final temperature. Care must be taken to comply with the temperature restrictions imposed by the stationary phase of the column, and also ensure that the temperature is appropriate for the volatility and complexity of the sample. The **GC–MS interface** temperature should be set to be at least as hot as the highest GC oven temperature to be used, and if possible 10 or 20 °C higher. Care should be taken to ensure that the GC column is not in direct contact with the walls of the GC oven, so avoiding any 'hotspots'.

GC–MS Operation and Analyses

The GC column should be conditioned before use, and this means heating the column in the GC oven with the helium gas flowing through up to the

Table 3.4 *Comparison of various GC injectors*

Injection mode	Sample type
Splitless: hot injector; initial oven *ca.* 10 °C lower than the boiling point of the solvent; 2 to 4 µL injection	All of sample is injected onto column; useful for dilute samples, and samples that contain heavy by-products
Split: hot injector; initial oven temp. is not critical; 1 µL injection	A percentage (typically 1 to 10%) of the sample is injected onto column; useful for concentrated samples
On-column: cold injector, initial oven ± 10 °C of the boiling point of the solvent; 0.5 to 2 µL injection	All of sample is injected onto column; useful for dilute, thermally sensitive samples, but not involatile ones

maximum recommended temperature for two or three hours if possible. This operation can be performed with the GC column in position in the ionization source.

In practice, the ionization source is optimized and calibrated as described for EI, CI+, or CI− analyses using reference and standard compounds introduced from either the reference inlet or by use of a direct insertion probe (Chapter 3, Sections 2 and 3).

A good way to test the operation of the GC and the GC–MS interface is to run a standard sample, and one of the most suitable is the **Grob test mixture**.[16] The Grob test mixture is available commercially, usually in solution in hexane or dichloromethane, and generally contains at least some of the following compounds: butane-2,3-diol, decane, undecane, dodecane, 2,6-dimethylaniline, 2,6-dimethylphenol, methyl decanoate, methyl undecanoate, methyl laurate, and octan-1-ol. From this list it can be seen that the compounds have been carefully chosen to include chromatographically 'difficult' compounds such as the acidic and basic components, and additionally adjacent members of homologous series. The Grob test mixture is suitable for most GC columns, although it is always best to check with the supplier's literature. After analysis, probably using a 2 µL splitless injection at an injector temperature of 250 °C and an initial oven temperature some 10 °C lower than the boiling point of the solvent, the oven temperature can be ramped at a rate of 20 °C min⁻¹ to 250 °C, and held at this temperature for 5 min, or until all the components have been detected. If 250 °C is the uppermost oven temperature, then the GC–MS interface should be held at a constant 250 to 270 °C throughout the analysis.

If a splitless injection is made, then all of the solvent and sample will pass onto the column and into the ionization source, and this will cause an enormous fluctuation in the operating pressure of the source when the solvent front arrives there. This could easily damage the filament, and in order to prevent this happening, the filament is not switched on until after the solvent

16 K. Grob, G. Grob and K. Grob, *J. Chromatogr.*, 1978, **156**, 1.

front has reached the ionization source. The time of arrival of the solvent front in the source depends on the length and film thickness of the GC column, but is usually in the region of 2–5 min after the injection has been made. The arrival time can be measured by injecting pure solvent through the GC with the filament switched off, and observing the source ionization gauge, which will swing markedly to higher pressures when the solvent appears in the source. Once this time has been measured, a 'solvent delay' time can be inserted into the acquisition parameters for future sample analyses.

Another matter to be aware of when acquiring GC–MS data is that the GC peaks eluting from a capillary column will be only a few seconds wide, and so a fast scan time should be set. As most compounds being analysed by GC–MS will have molecular weights of less than 800 da, the scan range to be covered is relatively small and scan speeds of 1 s or even faster should be sustainable on modern mass spectrometers.

After having analysed the Grob text mixture, with approximately 1 ng of each component going through to an electron impact ionization source, the data should be inspected carefully (see Figure 3.10). All of the chromatographic peaks should be symmetrical in shape and no tailing should be evident. The peaks should be well separated and the spectra from each one of a suitable quality for library searching. From the mass chromatogram, such GC parameters as the theoretical plate number (N) and the effective plate number (N_{eff}) can be calculated. One of the most useful and easy to calculate parameters is the **Trennzahl (TZ)** or **Separation Number**. To calculate this value, two adjacent members of an homologous series need to appear in the same chromatogram, and with the Grob mixture there are usually several possibilities for this, *e.g.*, decane and undecane, or methyl decanoate and methyl undecanoate. Once the components of interest have been identified, the separation number can be calculated by dividing the distance between the two peaks by the sum of their peak widths at half height, and then subtracting one from this number to give the final total in accordance with Equation (3.4). The number obtained should be approximately equal to the length of the column in metres, and the larger the separation number, the more efficient the total system,

$$TZ = \left(\frac{d}{a_{\frac{1}{2}} + b_{\frac{1}{2}}}\right) - 1 \tag{3.4}$$

where $a_{\frac{1}{2}}$ = the width of peak a at half height,
$b_{\frac{1}{2}}$ = the width of peak b at half height,
d = the distance between peaks a and b.

For the Grob test mixture illustrated in Figure 3.10, the peaks with retention times of 7.36 and 8.24 min correspond to the two methyl esters, methyl decanoate and methyl undecanoate. The separation number calculated from these two components was $[21 /(0.3 + 0.5)] - 1 = 25$. The length of the column

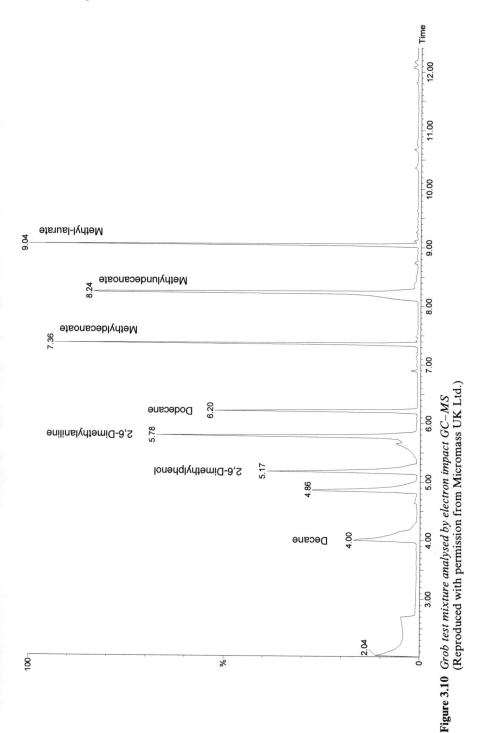

Figure 3.10 *Grob test mixture analysed by electron impact GC–MS* (Reproduced with permission from Micromass UK Ltd.)

Table 3.5 *Trouble-shooting guide to GC–MS*

Problem	Possible inference
1. Baseline drifts in a negative manner	carrier gas leak
2. Baseline drifts in a positive manner	column bleed contaminated column
3. Elevated baseline	column bleeding impure carrier gas
4. Noisy baseline	dirty injector dirty column detector temperature too high
5. Tailing peaks	dirty injector dead volume at the injector dirty liner septum needs replacing injector temperature too low GC–MS interface temperature too low
6. Leading peaks	column overloaded oven temperature too low
7. Ghost peaks	injector liner dirty septum needs replacing
8. Peak broadening	GC–MS interface temperature too low carrier gas flow too low
9. Septum peaks	septum needs replacing septum contaminating liner
10. Sample memory effects	dirty liner injector temperature too low
11. Column bleed	damaged column oxygen in carrier gas oven temperature too high interface too hot
12. Retention times non-reproducible	carrier gas leak GC flow/pressure control faulty

in this case was 25 m, so there is a good correlation between these two parameters, and this particular system was used with confidence for further sample analyses.

If the two numbers differ widely then steps must be taken to improve the chromatography and GC–MS coupling. A summary of possible GC–MS problems and some trouble-shooting suggestions are given in Table 3.5. The main areas for concern in GC–MS work are the injector, the GC–MS interface and the carrier gas.

The GC–MS interface temperature is important; if it is too low all the chromatographic peaks emerging will be too broad and will show signs of tailing, and if the interface temperature is too high then the baseline will become noisy and there will be evidence of excessive column bleed.

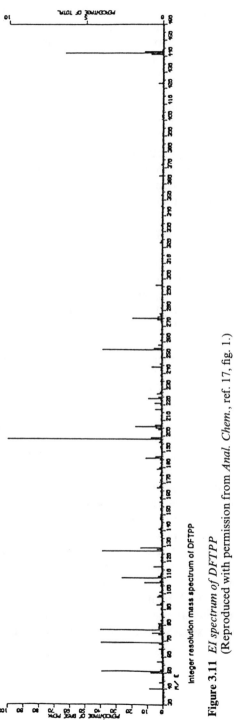

Figure 3.11 *EI spectrum of DFTPP*
(Reproduced with permission from *Anal. Chem.*, ref. 17, fig. 1.)

Injector problems may include: a dirty injector, which will show up as a noisy base-line; tailing chromatographic peaks, which can be as a result of a dirty injector or liner, a poor quality septum, or the injection temperature being too low; ghost peaks, due to a dirty liner or a septum that needs replacing; septum peaks, which indicate that the septum needs to be changed. The ions arising from the septum are usually similar to the ones arising from column bleed (*i.e.* m/z 207, 281, 355). The two problems can be distinguished, however, because the column bleed peaks will be present constantly as a background that gradually rises with increasing GC oven temperature, while septum-related material is flushed onto the GC column during the injection process and so appears as a chromatographic peak.

Carrier gas problems can result in the baseline drifting in a negative fashion, which indicates that the carrier gas is leaking, an elevated baseline implying that the carrier gas is contaminated, or broad peaks that may be sharpened if the carrier gas flow rate is increased.

It is worthwhile checking the following points before embarking on any analyses.

- The carrier gas should be checked for leaks, that the flow rate is set correctly, and that the gas is of a high purity.
- The injector should be fitted properly, be set at the correct temperature, and the septum and liner should be clean and in good condition.

Another standard compound worth a mention at this stage is decafluoro-triphenylphosphine (DFTPP), which is used as a GC–MS standard by the United States Environmental Protection Agency (EPA). DFTPP was chosen to calibrate, or standardize, the ion intensity scale from one mass spectrometer to another, so that data can be reproduced accurately on any number of different mass spectrometers.[17] The idea was to generate standard mass spectra, rather than mass spectra unique to any one particular instrument. Briefly, the procedure developed involves using a solution of DFTPP in acetone at a concentration of 10 ng μL^{-1} and analysing 2 μL of this solution through a standard GC column of the dimethylpolysiloxane type. Scanning over the range m/z 33–500 with a scan speed sufficiently fast to acquire at least four spectra throughout the chromatographic peak, an averaged spectrum (see Figure 3.11) should present the DFTPP related ions, with their intensities falling within the limits stipulated, as given in Table 3.6.

Although it is not necessary to keep to these guidelines during normal laboratory work unless the work is EPA related, it is sometimes useful to be able to set up standard conditions before undertaking analyses of unknowns, or if it is necessary to reproduce spectra from one mass spectrometer to another.

[17] J. W. Eichelberger, L. E. Harris and W. L. Budde, *Anal. Chem.*, 1975, **47**, 995.

Table 3.6 *EI ion abundance limits for DFTPP*[17]

DFTPP ions (m/z)	Ion abundance limits
51	30 to 60% of *m/z* 198
68	less than 2% of *m/z* 69
70	less than 2% of *m/z* 69
127	40 to 60% of *m/z* 198
197	less than 1% of *m/z* 198
198	base peak, 100% relative abundance
199	5 to 9% of *m/z* 198
275	10 to 30% of *m/z* 198
365	1% of *m/z* 198
441	less than *m/z* 443
442	greater than 40% of *m/z* 198
443	17 to 23% of *m/z* 442

Liquid Chromatography Using the Particle Beam Interface

The majority of liquid chromatography–mass spectrometry techniques are based on the so-called 'soft' ionization methods such as the atmospheric pressure ionization methods, together with the earlier techniques of thermospray and plasmaspray, thus providing molecular weight information. The high degree of fragmentation, and hence structural information and library search compatibility, available from electron impact makes Particle Beam EI/ CI a highly desirable method of ionization, although the difficulties in interfacing an ionization method that comprises the use of a reasonably fragile filament in a high vacuum with a chromatographic technique eluting *ca.* 1 mL min^{-1} are evident.

The pioneer workers in this field christened the technique MAGIC–LC–MS (monodisperse aerosol generation interface for combining liquid chromatography with mass spectrometry),[18] although the methodology termed particle beam (PB) has since evolved. Basically the mobile phase eluting from the LC column enters a heated chamber (*ca.* 40 to 50 °C) where it is nebulized (using helium gas) into a fine spray of droplets from which the volatile solvent molecules evaporate more readily than the solute molecules (Figure 3.12). The solute molecules tend to remain as particles which then pass through a nozzle into a chamber under vacuum, where further solvent molecules are removed. The desolvated solute molecules arrive in the EI or CI source as a spray of uncharged particles, and are then ionized by conventional EI or CI methods. Different particle beam interfaces will have varying numbers of and modifications to the chambers described here, but the basic principle will be the same. The recommended procedures should be read and followed faithfully to achieve optimum performance for any interface.

[18] P. C. Winkler, D. D. Perkins, W. K. Williams and R. F. Browner, *Anal. Chem.*, 1988, **60**, 489.

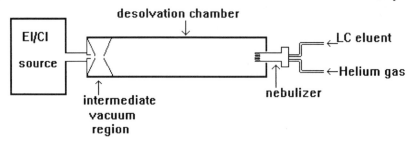

Figure 3.12 *Typical particle beam interface*

Particle beam interfaces generally cope with flow rates of between 0.5 and 1.5 mL min^{-1}, although the system is more tolerant to normal phase (organic) solvents than reverse (aqueous) based ones. Caffeine is a useful test sample with which to optimize the source and interface conditions, and the spectrum generated should be library searchable using the standard EI libraries available. For tuning purposes, a mobile phase of 1:1, (v/v) methanol/water is a good starting point, and as usual the optimum sensitivity of the system should be confirmed before sample analyses are undertaken. Any reduction in sensitivity will probably be as a result of source or interface contamination, and if this occurs, the first check to make is to establish the EI/CI source sensitivity by disconnecting the LC and nebulizing gas and admitting a standard such as heptacosa through the reference inlet. This will indicate whether the problem is within the source or the interface.

The source vacuum is of importance and a good vacuum must be maintained at all times. If the pressure in the source is too high, *i.e.* significantly higher than the pressure obtained under standard CI conditions, then the temperature of the desolvation chamber should be checked, as well as the helium flow, and it should be ensured that the extra backing pump for the intermediary vacuum stage is operating properly, and if necessary, under gas ballast conditions.

Particle beam has been used in many environmental areas, especially for the analysis and confirmation of pesticides,[19] although this technique is not very suitable for extremely volatile samples, which tend to be pumped away along with the solvent, or for thermally labile compounds because the source temperature is held at higher temperatures than normal, say 200 to 250 °C. Particle beam is still used to some degree, in conjunction with both electron impact and chemical ionization, although the API methods have largely superseded it wherever possible, due to their ease of use, reliability, and high sensitivity.

[19] T. D. Behymer, T. A. Bellar and W. L. Budde, *Anal. Chem.*, 1990, **62**, 1686.

Supercritical Fluid Chromatography

Supercritical fluid chromatography (SFC)[20] utilizes a supercritical fluid, which is similar in viscosity to a gas but closer in density to a liquid, as the mobile phase to elute samples through a chromatographic column. A supercritical fluid is obtained when the operating temperature is greater than the critical temperature of the fluid, thus enabling the fluid to have some solvating properties. As a result, SFC can be thought of as intermediate and hence complementary to both gas and liquid chromatography.

SFC does not require the sample to be volatilized, and so can be used for some thermally labile and non-volatile samples that are not amenable to GC. The chromatographic resolution is deemed to be greater than that observed with LC, and hence retention times can be shorter.

Both capillary column SFC and packed column SFC are used, and both have been interfaced successfully to mass spectrometry.[21] Packed column SFC employs LC columns, while capillary column SFC uses fused silica capillary columns. The flow rate of eluent associated with packed column SFC makes this technique more compatible with LC–MS interfaces such as thermospray (see Chapter 6, Section 3) and API,[22] and it is usually capillary column SFC which is used in conjunction with EI/CI ionization sources, although some modifications are necessary.[23] In simple terms, the end of the fused silica can be passed directly into the source because the flow rate of the eluent is compatible with maintaining the source vacuum, and although some interfaces are commercially available, a good many have been home built, as the technique has never reached the levels of exploitation that were anticipated when the initial studies were carried out. The reason for this lack of fame is due to several reasons, including a general feeling that SFC–MS is tricky and requires much maintenance, and that the more recent API ionization methods have proved reliable and readily automated.

However, the use of fused silica capillary column SFC provides good chromatographic resolution, albeit sometimes with long retention times.[24,25] If SFC–MS is being considered as a serious method for analyses, it is recommended that as much experience as possible is gained off-line before interfacing to a mass spectrometer, and that adequate caution is taken when using supercritical fluids.

One supercritical fluid that is often used is carbon dioxide, either alone[24,25]

[20] P. A. Peadon and M. L. Lee, *J. Liq. Chromatogr.*, 1982, **5**, 179.

[21] R. D. Smith and H. R. Usdaw, *Anal. Chem.*, 1983, **55**, 2266.

[22] J. F. Anacleto, L. Ramaley, R. K. Boyd, S. Pleasance, P. G. Sim and F. M. Benoit, *Rapid Commun. Mass Spectrom.*, 1991, **5**, 149.

[23] P. J. Arpino, J. Cousin and J. Higgins, *Trends Anal. Chem.*, 1987, **6**, (3), 69.

[24] D. E. Games, A. J. Berry, I. C. Mylchreest, J. R. Perkins and S. Pleasance, *Eur. Chromatogr. News*, 1987, **1**, 10.

[25] D. E. Games, A. J. Berry, I. C. Mylchreest, J. R. Perkins and S. Pleasance, *Lab. Pract.*, 1987, **2**, 45.

or with alcoholic modifiers such as 3–5% (*w/w*) propan-2-ol[26,27] to enhance the chromatography. The critical temperature of carbon dioxide is 31.05 °C and its critical pressure 72.9 atm, which are relatively easy to achieve in order to produce the supercritical fluid, and additionally carbon dioxide is not toxic or inflammable. However, because most work has been performed on individual systems, parameters such as column length, restriction and associated fittings, temperatures, *etc.* vary considerably, and it is worth checking with the literature for the precise conditions reported for a particular application. Separation temperatures of between 35 and 100 °C have been used. The fused silica capillary, usually 10 m in length, can be passed through the GC–MS interface, with a restrictor fitted in order to maintain the supercritical properties of the fluid within the column. Both EI and CI analyses have been performed, although the EI sensitivity is deemed to be less than under GC–MS conditions, and good quality data have been presented on polystyrene,[24,25] oligosaccharides,[26] and drugs and their metabolites,[28] *inter alia.* Microbore (1 mm) LC columns have also been coupled to EI/CI ionization sources.[29]

Calibration with a polysiloxane compound has been used for SFC analyses using ammonia CI as reagent gas in the CI ionization source.[27]

[26] J. D. Pinkston, G. D. Owens, L. J. Burkes, T. E. Delaney, D. S. Millington and D. A. Maltby, *Anal. Chem.*, 1988, **60**, 962.

[27] J. D. Pinkston, G. D. Owens and E. J. Petit, *Anal. Chem.*, 1989, **61**, 777.

[28] E. D. Lee, S-H. Hsu and J. D. Henion, *Anal. Chem.*, 1988, **60**, 1990.

[29] H. T. Kalinoski and R. D. Smith, *Anal. Chem.*, 1988, **60**, 529.

CHAPTER 4

Fast Atom/Ion Bombardment Ionization, Continuous Flow Fast Atom/Ion Bombardment Ionization

1 What type of Compounds can be Analysed by Fast Atom/Ion Bombardment Ionization Techniques?

Fast atom/ion bombardment (FA/IB) is well suited to organic compounds which have some degree of polarity, and which contain either acidic and/or basic functional groups. Those compounds which have basic groups tend to run well in positive ionization mode, and those with acidic centres give good results in negative ionization mode. Compounds that have both acidic and basic functionalities can be analysed by either positive or negative ionization. Other compounds which are amenable to FA/IB are those which are already charged, such as quaternary ammonium compounds with a positive charge, and alkali metal salts of carboxylic acids, which have a negative charge. Molecular weight information is the primary type of data obtained in both positive and negative ionization modes.

On this basis, it soon becomes evident that a great number and variety of compound classes are amenable to FA/IB, and a brief summary includes peptides,[1,2,3] proteins,[4,5,6,7] fatty acids,[8] organometallics,[9,10,11] surfactants,[11] carbohydrates,[12] antibiotics[13] and gangliosides.[14]

[1] M. Barber, R. S. Bordoli, R. D. Sedgwick and A. N. Tyler, *J. Chem. Soc., Chem. Commun.*, 1981, 325.
[2] M. Barber, R. S. Bordoli, G. J. Elliott, R. D. Sedgwick and A. N. Tyler, *Anal. Chem.*, 1982, **54**, (4), 651.
[3] L. C. E. Taylor, *Industrial Research and Development*, September 1981.
[4] M. Barber, R. S. Bordoli, R. D. Sedgwick, A. N. Tyler, G. V. Garner, D. B. Gordon, L. W. Tetler and R. C. Hider, *Biomed. Mass Spectrom.*, 1982, **9**, 265.
[5] M. Barber, R. S. Bordoli, G. J. Elliott, R. D. Sedgewick, A. N. Tyler and B. N. Green, *J. Chem. Soc., Chem. Commun.*, 1982, 936.
[6] W. Aberth, K. M. Straub and A. L. Burlingame, *Anal. Chem.*, 1982, **54**, 2029.
[7] M. Barber and B. N. Green, *Rapid Commun. Mass Spectrom.*, 1987, **5**, 80.
[8] K. B. Tomer, N. J. Jensen and M. L. Gross, *Anal. Chem.*, 1986, **58**, 2429.
[9] T. R. Sharp, M. R. White, J. F. Davis and P. J. Stang, *Org. Mass Spectrom.*, 1984, **19**, 107.

The use of FA/IB in conjunction with magnetic sector and quadrupole mass spectrometers is described in this Chapter.

2 Fast Atom/Ion Bombardment

The Principles of Fast Atom/Ion Bombardment Ionization

The majority of the early mass spectrometric ionization techniques, such as electron impact and chemical ionization, require the sample to be present in the ionization source *in its gaseous phase*. Volatilization of the sample is generally achieved by applying heat to the sample, which precludes several types of compound classes such as the thermally labile and the involatile, including highly polar samples and those of very high molecular mass. Chemical derivatization has been used in many cases to improve volatility and thermal stability, but still many compounds remained elusive until fast atom bombardment (FAB) emerged to remedy this situation.[1,2]

The FAB technique was first described[1,2,3] as using a 'simple sputter ion source'. A high velocity, rare gas atom molecular beam was produced in the ionization source, and directed onto the sample, which was in solution on a target, thus causing desorption of protonated or deprotonated molecular ions from the sample. These sample-related ions are analysed subsequently by the mass spectrometer (see Figure 4.1).

The fast atom beams most frequently used are generated from xenon or argon atoms. The production of a fast neutral beam usually takes place in an enclosed unit (often known as the 'gun' or 'ion gun') which is situated in close proximity to the ionization chamber. The process requires the formation of an initial ion beam (Xe^+) that can be accelerated to a known kinetic energy ($\underline{Xe}^+$) and focused into a beam of high intensity. This beam enters a chamber which has a pressure of *ca.* 10^{-3}–10^{-4} torr, and under these conditions a charge exchange process occurs, with little or no loss of forward momentum, and the result is a fast moving beam of neutral particles ($\underline{Xe}$) with a controllable kinetic energy of *ca.* 3–10 keV, see Equation (4.1),

$$Xe \rightarrow Xe^+ + e^-$$
$$Xe^+ \rightarrow \underline{Xe}^+$$
$$\underline{Xe}^+ + Xe \rightarrow \underline{Xe} + Xe^+ \qquad (4.1)$$

where fast moving particles are denoted by underlining.

[10] R. Davis, I. F. Groves, J. L. A. Durrant, R. Brookes and I. A. S. Lewis, *J. Organomet. Chem.*, 1983, **241**, C27.
[11] J. L. Gower, *Biomed. Mass Spectrom.*, 1985, **5**, 191.
[12] J. P. Kamerling, W. Heerma, J. F. G. Vliegenthart, B. N. Green, I. A. S. Lewis, G. Stecker and G. Spik, *Biomed. Mass Spectrom.*, 1983, **10**, 420.
[13] S. Santikarn, S. J. Hammond, D. H. Williams, A. Cornish and M. J. Waring, *J. Antibiot.*, 1983, **36**, 362.
[14] K. Vékey, *Org. Mass Spectrom.*, 1989, **24**, 183.

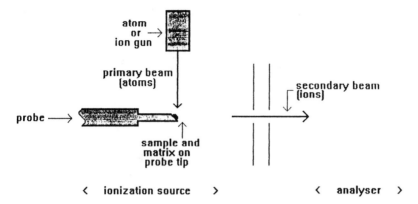

Figure 4.1 *Fast atom bombardment (FAB) ionization process*

In a typical operation, a fast moving xenon ion beam would be generated and then charge-exchanged to produce a fast moving xenon atom beam. Once the fast moving neutral beam has been formed, it cannot be focused in the same way that an ion beam can, and so the alignment of the fast atom beam onto the sample-holding target is of great importance. The optimum angle of incidence of the fast atom beam onto the sample was found to be $20°^2$ (see Figure 4.2).

Some of the momentum of the fast moving atom beam is transferred on collision with the sample and matrix (a relatively involatile solvent which holds the sample in position on the sample target) molecules, producing positive and negative sample-related ions which are ejected from the sample/matrix surface and can then be extracted and accelerated into the mass analyser. Generally FAB produces protonated, MH^+, or deprotonated, $(M-H)^-$, molecular ions by means of a chemical ionization process in which the matrix can be regarded as the chemical reagent. These *quasi*-molecular ions have a little excess energy and will sometimes produce fragment ions of somewhat lower intensity. FAB is classified as a mild or soft ionization technique which produces primarily molecular weight information. Positive and negative ionization mass spectra are produced with equal facility. Although FAB was first used with magnetic sector mass spectrometers, the technique was transferred painlessly to quadrupole mass spectrometers.

Once set up, FAB is usually routine and straightforward. The energy of the atom beam is controlled by the high voltage applied to the gun. The intensity is regulated by adjusting the gas flow. The choice of gas has generally been one of the inert gases, and helium (atomic mass 4 da), argon (atomic mass 40 da), and xenon (atomic mass 131 da) have all been used successfully. The higher mass atoms tend to be preferred, with xenon being the most popular despite its high cost, as these generate the maximum yield of secondary ions.

The matrix plays an important role in this technique. It is necessary to have

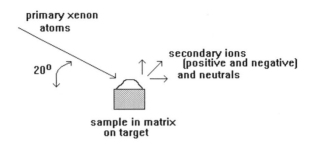

Figure 4.2 *Optimum angle of incidence of the fast atom bombardment ion beam onto the sample*

a matrix present in order to transport the sample from the laboratory bench into the high vacuum environment of the source of the mass spectrometer, and keep the sample in position on the probe tip until a mass spectrum has been acquired. Therefore the matrix must be reasonably involatile and have good solvating properties for the type of sample under investigation, which generally means a liquid at room temperature and atmospheric pressure that does not evaporate too quickly in the vacuum of the mass spectrometer. In order to meet with these requirements, the matrices are usually quite polar substances with molecular weights of, say, less than 300 da, and so produce distinctive mass spectra themselves, often consisting of polymeric units, or clusters $\{[MH + (M)_n]^+$ or $[(M-H) + M_n]^-\}$ in addition to their protonated or deprotonated molecular ions $[(MH)^+$ or $(M-H)^-]$. More discussion about specific matrices can be found later in this section.

FAB, when first announced, was remarkable in that molecular masses for small proteins could be generated and even as early as 1982, glucagon (MW 3481 da) and the oxidized β-chain of bovine insulin (MW 3494 da) had been analysed,[4] followed closely by bovine insulin (MW 5729.6 da).[5] (It should be noted that all these results were obtained on double focusing magnetic sector mass spectrometers.) It soon became apparent, however, that for proteins with molecular masses above 10 kda, higher ion yields from these samples were necessary.

It was for this purpose that a caesium (Cs^+) ion gun[6] was developed and it was shown that a fast beam of ions could be used to bombard the sample target with higher energies than a fast beam of atoms. In fact the Cs^+ ion beams used generally have ion energies of up to 35 kV, compared with the fast atom beam energies of 8–10 kV. The name fast ion bombardment (FIB) was coined to distinguish between the two methods of generating ions, although FIB is referred to more deservedly as liquid secondary ion mass spectrometry (LSIMS) due to the fact that both the primary and the secondary beams are composed of ions. The term 'liquid' is present to denote the fact that the sample is dissolved in a liquid matrix.

The caesium gun consists of a filament which is used to heat a caesium pellet from which caesium ions are then evaporated (see Figure 4.3). The caesium

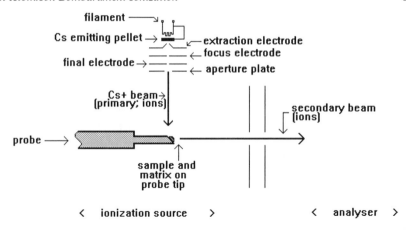

Figure 4.3 *Fast ion bombardment (FIB) or liquid secondary ionization mass spectrometry (LSIMS) ionization process*

ions are accelerated and extracted from the caesium gun and subsequently focused onto the target at the end of the probe on which the sample is placed after dissolution in the matrix. The substrate and matrix ions produced (which together constitute the secondary ion beam) are extracted and analysed by the mass spectrometer. The energy of impact of a caesium ion when it reaches the target is variable, sometimes up to 35 kV, and can be optimized for any particular sample. The spectra generated are usually indistinguishable in nature from FAB spectra, especially if working with low molecular weight samples (< 1000 da) at reasonable concentrations. However, the high energy caesium ion beam *does* increase the sensitivity of the technique with all samples in general and the results are most striking with higher molecular mass compounds. Aided by this extra sensitivity, molecular weight data have been published (again with a double focusing magnetic sector mass spectrometer) on proteins such as trypsinogen, of molecular mass 23 978 da.[7]

Practical Operation

Requirements for Operation

Most mass spectrometers which have the capability of analysing samples by FAB or FIB/LSIMS tend to be fitted with either one alternative or the other, and such mass spectrometers have usually either a magnetic sector or a quadrupole analyser, and often hybrid or tandem combinations of these two. The scope of both FAB and FIB/LSIMS covers basically the same type of compounds, although the sensitivity (and hence sample mass range) is greater with FIB/LSIMS. Both variations of the technique produce results quickly and often little expertise is required to generate data. In order to perform FAB or FIB/LSIMS the following equipment is necessary:

- a mass spectrometer, with both the source and analyser regions under vacuum, equipped with a FAB or FIB/LSIMS source, and a caesium ion or fast atom bombardment gun with an appropriate power supply;
- a direct insertion FAB or FIB/LSIMS probe, with at least one, and preferably two or three, clean, interchangeable probe tips;
- if FAB is being used, then a small cylinder of xenon or argon;
- several suitable matrices and additives (see later in this Section);
- concentrated hydrochloric acid (for washing the probe tip);
- suitable pipettes, syringes, or glass capillaries for applying the matrices and sample solutions to the probe tip;
- suitable sample vials for mixing the samples, solvents, matrices, and additives.

Setting up and Using Fast Atom/Ion Bombardment Ionization Mass Spectrometry with the FAB Direct Insertion Probe

As outlined, FAB and FIB/LSIMS both require a working mass spectrometer equipped with a FAB or FIB/LSIMS source and either a fast atom or a caesium ion gun. The mass spectrometer must have attained a suitable, working vacuum before the direct insertion probe is introduced into the source.

For FAB operation, the inert gas must be connected to the fast atom gun. Xenon is the gas most frequently used. Before switching on the fast atom gun, the gas line should be flushed out once or twice with the inert gas to make sure that all the lines are filled with the gas, and all of the air previously in the lines has been eliminated. This will prolong the lifetime of the fast atom bombardment gun. Once this has been achieved, the fast atom gun can be switched on and off as required. It is often more convenient to leave a small cylinder of the inert gas permanently connected to the fast atom gun on the mass spectrometer, rather than connecting and disconnecting the gas lines every time FAB is used. This saves the relatively expensive inert gas for the ionization process rather than using copious amounts for flushing out the gas lines.

The intensity of the atom beam is controlled by regulating the gas flow through the gun, and a typical gas flow of *ca.* 0.5 mL min^{-1} should be set. Once the gas flow has been set, it should require only minor further adjustment. When the atom gun is switched on a voltage of *ca.* 8 kV should be obtained. The readings on the FAB gun control supply should not fluctuate once set, and should be compatible with the expected operational parameters suggested in the appropriate instruction manual. No further tuning is required for the FAB gun.

For FIB/LSIMS operation, the caesium gun should be switched on, the high voltage raised in increments to its operating level, and the readbacks noted. Again the manual should be consulted for the correct method of operation and suggested operational parameters. Further tuning when low concentration levels of samples or high molecular mass samples are under scrutiny is required for optimum performance. *Appropriate care should always be exercised when using high voltages.*

These techniques are best tested with a known sample applied in a known matrix, and this test sample can be anything that has been analysed successfully previously. Traditionally FAB and FIB/LSIMS have been used more for peptides than for any other type of compound, and so a well-documented peptide such as one of the angiotensins, bradykinins, enkephalins, or Substance P should suffice. For test purposes, a solution of one of these small peptides in water or methanol containing 1% formic acid at a concentration of $10 \ \mu g \ \mu L^{-1}$ should be prepared. A 1 μL aliquot of this solution should then be added to and mixed with the matrix, which has already been placed on the probe tip. The matrix for these purposes can be glycerol [$HOCH_2CH(OH)$-CH_2OH; MW 92.05 da] and *ca.* 2 μL should have been spotted onto the *clean* probe tip.

Probe tips come in many different sizes and shapes, and so the volume of matrix required to cover the probe tip will vary from one probe tip to another. All of the surface should be covered however, not just one half, and the probe tip needs to be properly 'wetted'. This means that the matrix should not sit as a small blob on one part of the probe tip, but should be spread out evenly over the entire surface area. If this is not the case then the probe tip should be carefully and gently scratched using the tip of a glass pipette, and only when the matrix is completely covering the tip should the sample solution be added.

The material of the probe tip will also vary from one manufacturer to another, with stainless steel, copper, and gold-plated copper being the most commonly supplied. If the probe tip is dirty or stained, then it should be cleaned. First a range of solvents should be tried, but if the residual solids are resistant to these, the tip may be treated with 2 μL of concentrated hydrochloric acid, and then washed well with water. This usually removes even the most stubborn remnants of samples. Care should be taken with gold-plated probe tips which may well lose their gold-plate with too vigorous a cleaning.

So, with the direct insertion probe loaded with sample and matrix, and the high voltage of the FAB or FIB/LSIMS gun momentarily switched **off**, the probe can be inserted through the probe insertion lock. (If a magnetic sector mass spectrometer is being used, the high voltages of the source are often switched off when inserting the probe.) At this intermediate pumping stage, most of the solvent used to dissolve the sample initially (water or methanol) will be evaporated, leaving only the sample and the matrix on the target by the time the probe is inside the ionization source. Now the high voltage of the atom or ion gun should be switched **on** again (and the source voltages too, if these had been switched off), and sample and matrix ions should be readily visible on the oscilloscope or data system terminal. If a caesium ion beam is in use then further tuning may be necessary to optimize the sample ions, and sample-related ions such as the MH^+ or $(M-H)^-$ should be chosen for this purpose. The small peptides mentioned as being useful for test purposes will all produce MH^+ ions, and hence can be characterized in the positive ionization mode of operation.

The matrix, being a polar, organic compound will also produce a characteristic spectrum which will quickly become familiar, if not tedious, and which is comprised not only of MH^+ or $(M-H)^-$ ions, but also of polymeric cluster ions. Glycerol, for example, gives rise to ions at m/z 93 (MH^+), 185 [MH + M]$^+$, 277 [MH + (M)$_2$]$^+$, 369 [MH + (M)$_3$]$^+$ and so on up to high masses in positive ionization mode, and a complementary series starting with m/z 91 $(M-H)^-$, 183 [(M$-$H) + M]$^-$, 275 [(M$-$H) + M$_2$]$^-$, 367 [(M$-$H) + M$_3$]$^-$, and so on in negative ionization mode. A positive ionization FAB spectrum of glycerol is shown in Figure 4.4, and this spectrum is frequently superimposed on the spectrum of the sample. If, by chance, the sample gives rise to ions at any one of these m/z values, then the matrix should be changed.

Once the operator is satisfied that sample-related ions are being observed, then a spectrum should be acquired. Data should always be acquired at the highest possible resolution at which a reasonably intense signal is observed, as FAB and FIB/LSIMS produce singly charged ions and so a good resolution will not only generate a clear isotope pattern but also an unambiguous, monoisotopic molecular mass. For samples with molecular masses above *ca.* 10 000 da, if using a magnetic sector analyser, or *ca.* 2000 da, if using a quadrupole analyser, the signal strength will probably be too weak to permit acquisition at unit resolution, and low resolution data should then be acquired to generate *average*, rather than monoisotopic, molecular masses. The positive ionization FAB spectrum of methionine enkephalin, complete with matrix ions which could have been subtracted out, is shown in Figure 4.5.

As with all ionization techniques, the data produced are only as accurate as the calibration which has been employed to calculate their values. Calibration in FAB and FIB/LSIMS is readily accomplished in either positive or negative ionization modes with an aqueous solution of caesium iodide[14] at a concentration of 1 mg mL^{-1}, or various other metal halide solutions. Caesium iodide produces not only Cs^+ ions at m/z 133, but also a series of cluster ions [Cs + (CsI)$_n$]$^+$, where n = 1,2,3, *etc.* throughout the m/z range in positive ionization mode, and a complementary series of negatively charged ions in negative ionization mode, starting with I^- at m/z 127, and continuing with [I + (CsI)$_n$]$^-$ cluster ions (see Appendix 2). This calibrant has been used successfully up to 24 kda.[7,15]

If at all possible, it is advisable to keep one probe tip just for calibration purposes, in which case the possibility of sample contamination is avoided.

When a sample analysis has been completed, the high voltage of the FAB or FIB/LSIMS gun should be switched **off** and the probe removed.

[15] A. E. Ashcroft, A. D. Coles, S. Evans, D. J. Milton and B. Wright, *The Analysis of Peptides and Proteins by Mass Spectrometry*, Proc. Fourth Texas Symposium Mass Spectrom., College Station, April 1988, John Wiley & Sons, Chichester, UK, 1988, p. 279.

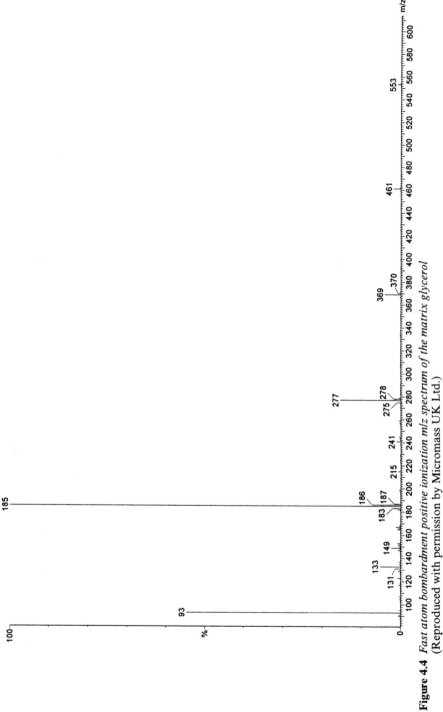

Figure 4.4 *Fast atom bombardment positive ionization m/z spectrum of the matrix glycerol* (Reproduced with permission by Micromass UK Ltd.)

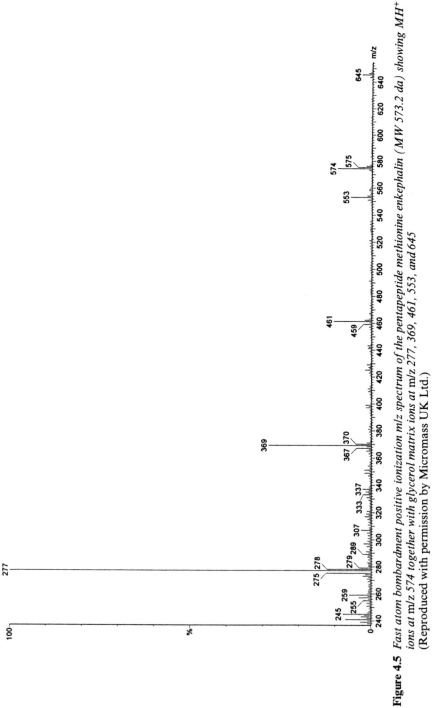

Figure 4.5 *Fast atom bombardment positive ionization m/z spectrum of the pentapeptide methionine enkephalin (MW 573.2 da) showing MH⁺ ions at m/z 574 together with glycerol matrix ions at m/z 277, 369, 461, 553, and 645* (Reproduced with permission by Micromass UK Ltd.)

The Analysis of Samples

Choice of Matrix

The choice of matrix is sample related, and does not depend on whether the bombarding beam is composed of atoms or ions. Although a multitude of matrices has been reported,[11] the majority of samples can be analysed with one or more of just a handful of these.

The purpose of the matrix is not only to transport the sample into the high vacuum region of the source of the mass spectrometer, but also it appears to be necessary for the ion formation process. The signal obtained from sample-related ions decreases as the matrix compound is depleted, but often can be regenerated simply by adding more matrix (*not* more sample) to the probe tip. Therefore the matrix should be capable of dissolving the sample, thus enabling sample molecules to sit on its liquid surface for maximum exposure to the bombarding beam. The matrix must be relatively involatile in mass spectrometric terms, in order to survive the high vacuum conditions. The more volatile the matrix, the shorter its lifetime on the probe tip, and so the shorter the signal from the sample ions. In order to compensate for this, and to permit the use of more volatile matrices, a cooled, direct insertion probe has been reported for LSIMS usage.[16]

The most common matrices *only* have been summarized in Table 4.1, together with their molecular formulae and masses, their most frequently encountered *m/z* ions, and their uses. Most FAB or FIB/LSIMS samples will produce acceptable spectra with one or more of these. For further information, the physical properties of a comprehensive list of FAB and FIB/LSIMS matrices have been tabulated.[17]

The matrix is often mixed with small quantities of an additive to improve the results. In the simplest case, this involves adding a small percentage (1–5%) of an acid to enhance protonation in the positive ionization mode, or a basic compound to aid deprotonation in the negative ion mode. More complicated cocktails have been published for very specific examples and tend to make FAB and FIB/LSIMS look like witchcraft! There are many references in the literature for different matrices and additives, providing the analyst with an almost endless list of suggestions to try for stubborn samples, including the few examples of matrix–additive complexes summarized here.

---

[16] A. M. Falick, F. C. Walls and R. A. Laine, *Anal. Biochem.*, *1986*, **159**, 132.
[17] K. D. Cook, P. J. Todd and D. H. Friar, *Biomed. Environ. Mass Spectrom.*, 1989, **18**, 492.

Table 4.1 *Some common FAB and FIB/LSIMS matrix compounds*

Matrix	Formula	Molecular weight	Associated ions (m/z)	Uses
glycerol	$C_3H_8O_3$	92.05	$[93 + (92)_n]^+$ $[91 + (92)_n]^-$	general purpose, with or without additives[11]
thioglycerol	$C_3H_8O_2S$	108.02	$[109 + (108)_n]^+$ $[107 + (108)_n]^-$	peptides, with 1% acetic acid added[5]
triethanolamine	$C_6H_{15}O_3N$	149.10	150 $(MH)^+$	fatty acids,[18] gangliosides[11] and anionic surfactants[11]
crown ethers, *e.g.* 18-crown-6	$C_{12}H_{24}O_6$	264.16	265 $(MH)^+$	organometallic complexes[9,11]
dithiothreitol/ dithioerythritol 3:1, 'magic bullet'	$C_4H_{10}O_2S_2$ $C_4H_{10}O_2S_2$	154.01 154.01	155 $(MH)^+$	organometallic complexes[11]
m-nitrobenzyl alcohol, 'MNBA'	$C_7H_7O_3N$	153.04	154 $(MH)^+$	general[19]

The addition of a number of acids to either glycerol or thioglycerol has been reported. Inorganic acids such as hydrochloric acid and sulfuric acid,[20] and organic acids such as acetic acid,[21] *p*-toluenesulfonic acid,[2] and oxalic acid[22] tend to increase the intensity of peptide spectra. The use of glycerol with co-solvents including dimethyl sulfoxide, dimethylformamide, alcohols, and water for the analysis of a whole range of compound classes has also been widely published.[11]

For an example of a very specific analysis, FAB ionization of chlorophyll can be examined. Although chlorophyll is totally insoluble in glycerol alone and hence does not produce a mass spectrum by this method, the addition of 1% of Triton X100 to the glycerol brought about solubilization to yield a green solution which produced both a positive and negative ion spectrum verifying the molecular weight (892 da) of chlorophyll.[2]

Finally, matrix-free FAB has been reported as producing improved results for one particular peptide compared to data obtained in the presence of a matrix.[23]

[18] K. A. Caldwell and M. L. Gross, *Anal. Chem.*, 1989, **61**, 494.

[19] J. Meili and J. Seibl, *Org. Mass Spectrom.*, 1984, **19**, 581.

[20] S. Naylor and G. Moneti, *Biomed. Environ. Mass Spectrom.*, 1989, **18**, 405.

[21] S. A. Martin, C. E. Costello and K. Biemann, *Anal. Chem.*, 1981, **54**, 2362.

[22] S. J. Gaskell, M. H. Reilly and C. J. Porter, *Rapid Commun. Mass Spectrom.*, 1988, **2**, (7), 142.

[23] L. C. E. Taylor, D. A. Brent and J. S. Cottrell, *Biochem. Biophys. Res. Commun.*, 1987, **145**, (1), 542.

Interpretation of Spectra

As discussed previously, positive ionization FAB and FIB/LSIMS produces protonated molecular ions while negative ionization generates deprotonated molecular ions, usually with little fragmentation, and so the spectra may be expected to be readily interpretable. If any fragmentation does occur, it can generally be rationalized from the chemical structure of the sample.

However, the presence of the matrix and any residual traces of solvent do make the spectra more complicated. Not only will the matrix protonate or deprotonate, it may also form cluster ions at regular intervals throughout the spectrum which may well coincide with the sample ions. If there are any traces of metal salts associated with the sample (*e.g.* with peptides and proteins) or with any of the solvents or additives which have been used, then both the sample and the matrix can give rise to either $(M + \text{metal cation})^+$ or $(M + \text{metal cation} - H)^-$ ions in positive or negative mode respectively. Sometimes more than one metal cation can add to a sample ion. The most commonly encountered adducts of this type are sodium (Na) or potassium (K), which add 23 or 39 respectively to the molecular mass of a sample. Ammonium adducts are common non-metallic adducts appearing at m/z MNH_4^+, some 18 amu higher than the molecular mass. Additionally sample–matrix adducts can be detected.

So if glycerol was the matrix being used, and a trace of a sodium salt was present, then in addition to the series of ions m/z 93, 185, 277, 369, *etc.* observed in positive ionization, a second series could replace or co-exist with this one, with ions occurring at m/z 115, 207, 299, 391, *etc.*

The sample may also exhibit both MH^+ and MNa^+ ions. Sometimes this can be useful in the interpretation of a spectrum, as a doublet of ions with a difference of 22 da indicates a sodium adduct and reveals the molecular weight of a sample with extra confidence.

Often metal cation adducts are formed consciously to aid interpretation of a spectrum. For example, if a sample produces a weak spectrum with several intense ions, none of which is particularly dominant, then 'doping' the sample with a trace amount of sodium or potassium chloride can produce additional ions, or increase the intensity of some of the exisiting ions, which may lead to a positive deduction of the sample's molecular weight.

If there is only one dominant ion in the spectrum, this could pose the dilemma of whether it is MH^+ or, say, MNa^+. In such a case the sample could be doped with a small percentage of a potassium salt. If, after addition of the potassium salt, a m/z shift of 38 da was observed, the original ion could be identified as MH^+, whereas a m/z shift of 16 da would indicate that the original ion was more probably MNa^+.

Limitations of FAB

With FAB and FIB/LSIMS, spectra containing valuable molecular mass information can be generated rapidly from a wide range of compound classes, including samples of relatively high mass. There are some limitations, however, of which the analyst should be aware. As discussed previously, the presence of the matrix gives rise to a high level of matrix-related ions. FAB and FIB/ LSIMS also generate a high chemical background which makes the spectra noisy below *m/z* 200, and so any sample-related ions in this region can be quite difficult to detect with confidence.

As with all ionization techniques that use a direct insertion probe, accurate and precise quantification is difficult to achieve. With FAB and FIB/LSIMS the sample signal often dies away when the matrix, rather than the sample, is consumed and so the operator cannot be sure that the ion signal obtained represents consumption of all the sample. Sometimes, a minor adjustment of the probe, either by rotating or changing its position within the source, will prolong the sample ion current.

Different samples can have very different responses to the FAB techniques, and ion signals can vary significantly for a given concentration of sample. One example of this is the analysis of peptide and protein mixtures, which has been a topic of some discussion.[24] The overall hydrophobicity or hydrophilicity of a peptide can be calculated from the Bull and Breese[25] indices of its amino acid constituents, and the more hydrophobic samples sit at the upper surface of the matrix where they are ideally placed for bombardment by the atom or ion beam and so produce a stronger signal than the more hydrophilic samples, which prefer to remain below the surface of the matrix. Hence the analysis of mixtures can present difficulties. One component of a mixture can suppress other components, and spectra that are consistent with a single-component sample may well have been obtained from a mixture. The component detected may not even be the major component. The analyst should be aware of this when investigating mixtures, for example protein digests,[26] when some components may remain undetected.

Despite these limitations, it must be acknowledged willingly that the introduction of FAB was a milestone in the development of mass spectrometric ionization techniques, opening up the rather narrow range of compound classes which were previously suitable for analysis, and solving many biochemical-related problems. It was and still is a popular technique to use, although it is now declining in use since the advent of electrospray.

3 Continuous Flow Fast Atom Bombardment

A Description of Continuous Flow and Frit FAB

Continuous flow, or frit, fast atom/ion bombardment offers a means of introducing samples in solution into a continuous flow of solvent which

[24] K. L. Busch, in 'Sample Peparation and Matrix Selection for Analysis of Peptides by FAB and LSIMS', *Mass Spectrometry of Peptides*, ed. D. M. Desiderio, CRC Press, Florida, ch. 10, 1991.
[25] H. B. Bull and K. Breese, *Arch. Biochem. Biophys.*, 1974, **161**, 665.
[26] S. Naylor, A. F. Findeis, B. W. Gibson and D. H. Williams, *J. Am. Chem. Soc.*, 1986, **108**, 6359.

terminates at the probe tip of a modified FA/IB probe. There is no need for removal and cleaning of the probe in between every sample analysis; the probe can remain in the ionization source all day while samples are injected through a conventional HPLC injection valve or solutions are simply drawn in by the high vacuum in the ionization source of the mass spectrometer. These techniques are particularly amenable to coupling with HPLC columns. There are design differences between frit FA/IB[27] and continuous flow (or dynamic) FA/IB,[28] although both use a hollow direct insertion probe of the same dimensions as the standard FAB probe. The modes of operation and usage are very similar, and the ionization of the sample is unchanged from that described for conventional FAB and FIB/LSIMS. To all intents and purposes, with continuous flow or frit FA/IB, it is simply the method of sample introduction that is different, and most points covered in this discussion will apply equally well to both methods.

Frit FA/IB[27] utilizes a length of fused silica tubing (typically 40 μm i.d.) which extends from the solvent reservoir or the outlet of the HPLC column to the tip of the FAB probe, and is contained within a stainless steel capillary inside the probe. At the probe tip, the silica tubing terminates flush with a stainless steel frit, and hence the name frit FA/IB. The first of these probes operated with a flow rate of just 0.52 μL min^{-1} and was coupled to an HPLC column of 0.26 mm internal diameter.

Continuous flow FA/IB[28] also involves the use of a fused silica capillary (typically 75 μm i.d.) which passes from a conventional injection valve, through the centre of the FAB probe to the probe tip. Later developments included the incorporation of a tissue wick at the probe tip to absorb excess solvent.[29]

The different methods of operation of continuous flow FA/IB are summarized in Figure 4.6, and involve:

(a) continuous sampling of the compound or reaction mixture of interest;
(b) consecutive, discrete injections of different samples; or
(c) coupling to a separation technique such as HPLC or capillary electrophoresis (CE).

The advantages of continuous flow FAB over conventional FAB include improved sensitivity and reduced background signal due to the fact that the matrix/sample ratio is rather lower, more accurate quantification because if the sample is injected through a conventional HPLC injector then it appears in the ionization source as a peak whose area can be measured, and more flexibility in operation and coupling to separation methods.

[27] Y. Ito, T. Takeuchi, D. Ishii and M. Goto, *J. Chromatogr.*, 1985, **346**, 161.
[28] R. M. Caprioli, T. Fan and J. S. Cottrell, *Anal. Chem.*, 1986, **58**, 2949.
[29] J. A. Page, M. T. Beer and R. Lauber, *J. Chromatogr.*, 1989, **474**, 51.

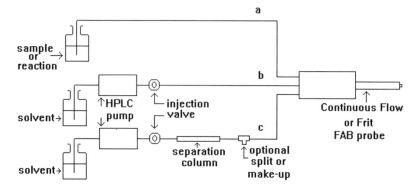

Figure 4.6 *Continuous flow fast atom/ion bombardment: methods of operation*

Practical Operation

Requirements for Operation

The following list describes the requirements for analysing samples with continuous flow FA/IB:

- a mass spectrometer, with both the source and analyser regions under vacuum, equipped with a FAB or FIB/LSIMS source, and a caesium ion or fast atom bombardment gun with a power supply;
- a continuous flow or frit FAB or FIB/LSIMS probe, with an intact length of fused silica capillary installed according to the manufacturer's instructions;
- if FAB is being used, then a small cylinder of xenon or argon;
- several suitable solvents, matrices and additives (see Section 4);
- optional HPLC injection valve and compatible syringe, HPLC solvent delivery pump, HPLC columns, fittings and ferrules.

Setting up and Using Continuous Flow Fast Atom/Ion Bombardment Ionization

Before commencing any dynamic analyses, it is advisable to check that static FA/IB is working properly, and to do this a known sample can be analysed as described in Section 2 of Chapter 4 using the standard, direct insertion probe. This ensures that the FAB or FIB/LSIMS gun is operating correctly, and also that the mass spectrometer is functioning. It also provides the analyst with the opportunity for calibrating the mass spectrometer over the appropriate m/z range, if this is necessary.

There are several general points to consider when setting up continuous flow FA/IB. It is critical to obtain and maintain a steady sample/matrix/

solvent (*i.e.* secondary, if LSIMS is the primary ion beam) ion beam, and for this purpose the continuous flow probe should be inserted into the mass spectrometer with a solvent flowing through it at a rate of 3 to 5 μL min^{-1}. The solvent flow can be generated by a pulse-free, solvent delivery pump, or simply by dipping the end of the fused silica into a solvent reservoir and relying on the source vacuum to 'pull' the solvent through. To start with, the solvent should be composed of 47.5:47.5:5 (*v/v/v*) water–acetonitrile or methanol–glycerol and a suitable matrix ion (*e.g. m/z* 93, 185, or 277 for glycerol in positive ionization) should be displayed and monitored on the oscilloscope or data system terminal. There may well be some instability to start with, and this can be observed both by watching the peak height and shape of the ions selected, and also the source vacuum gauge which will rise and fall suddenly with any fluctuations in the source's vacuum caused by solvent surges. A number of factors contribute to the stability of this ion beam, and these must be adjusted when necessary. To start with, the probe must be assembled properly, with the correct length of silica capillary. Usually 1 m is considered appropriate, as this is sufficient to maintain the high vacuum in the source, and to insulate the operator and HPLC equipment from the high voltage sources on magnetic sector mass spectrometers. The position of the silica at the probe tip should be checked; the silica should protrude 1 to 2 mm.

The solvent flow rate is critical, too low a flow will cause the liquid emerging from the probe tip to sputter rather than flow, and too high a flow rate will upset the vacuum in the ionization source and result in an unstable ion beam. The full range of flow rates is 1 to 20 μL min^{-1}. Solvents employed are usually, but not necessarily, of the reversed phase nature *i.e.* water, methanol, acetonitrile and combinations thereof, because of the (usually) high polarity of the samples under investigation. Some normal phase solvents such as toluene have also been used successfully. A small percentage, 1 to 10% (*v/v*) is optimum, of the solvent should be a matrix material, as this is still required to provide stability of evaporation in the ionization source and for ionization of the sample. The choice of matrix is not so critical with continuous flow FA/IB because far less matrix, relative to the amount of sample, is used. Glycerol is the popular choice.

The temperature of the probe tip must also be set correctly. With standard FA/IB, the ionization source is not heated, and this is a beneficial feature of the technique in the analysis of thermally labile samples. However, continuous flow FA/IB delivers a constant flow of solvent and matrix to the source, and gentle heat is now necessary to help remove excess solvent, and to prevent excessive cooling brought about by the evaporation of solvents. The temperature required at the probe tip depends on the solvent composition and volatility, and its flow rate; 40 °C is a good starting point, but the optimum temperature will vary from one mass spectrometer to another. A stable ion beam may take some time to achieve, but it is time well spent and necessary for uninterrupted reproducible analyses.

Having obtained a stable ion beam, the test sample chosen for static FA/IB

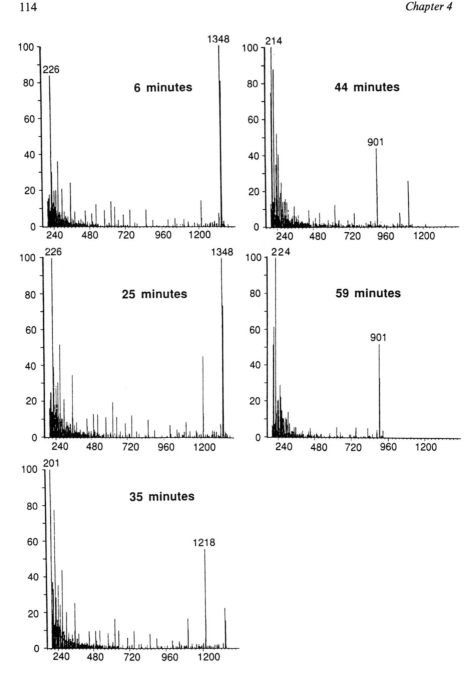

Figure 4.7 *On-line reaction monitoring of the carboxypeptidase Y enzyme digest of the peptide substance P by continuous flow fast atom bombardment* (Reproduced with permission from Kratos Analytical.)

can be repeated for dynamic FA/IB. The amount of sample required to produce a recognizable signal for continuous flow FA/IB will depend on the mass spectrometer in use, but the amount of sample required for standard FA/IB should now be known, and this is a good level to start at with continuous flow FA/IB. The sensitivity should be somewhat improved, and it should be possible to decrease the sample concentration if continuous flow FA/IB is operating efficiently.

Now the mode of analysis can be set up. If continuous infusion is the method of choice [Figure 4.6(a)] then the equipment needed is minimal. Quite simply, the fused silica capillary is fed into the sample-containing vial, and the high vacuum of the mass spectrometer is relied on to draw the solute–solvent mixture into the ionization source. For a fused silica capillary of approximately 1 m in length, and a source under high vacuum, this will result in a *ca.* 1 to 3 μL min⁻¹ flow rate. The flow obtained in this manner is usually pulse-free and quite stable, although it will take several minutes for the solute–solvent mixture to reach the probe tip. This method has been used for monitoring enzyme digest reactions[30,31] and the *in vivo* metabolism of drugs.[31] On-line reaction monitoring of the carboxypeptidase Y enzyme digest of the peptide Substance P (amino acid sequence ArgProLysProGlnGlnPhePheGlyLeuMet) by continuous flow FAB[32] is illustrated by the series of spectra taken at different times throughout the reaction (see Figure 4.7). Initially the MH⁺ ions can be seen at *m/z* 1348, while the ions at *m/z* 1218 correspond to the fragment remaining after cleavage of the first amino acid from the carboxy terminus, a methionine residue. Sequential losses of leucine, glycine, and phenylalanine leave residues which are consistent with the ions observed at *m/z* 1105, 1048, and 901 respectively. The use of a matrix is still necessary for this type of analysis.

The second mode of operation, depicted in Figure 4.6(b), illustrates the use of the continuous flow probe with a standard HPLC solvent delivery pump and injection valve. In this way a number of discrete samples can be analysed sequentially by injection through the valve, while the solvent is set to flow continuously into the source. There is no need to remove the probe in between analyses as the solvent will wash away all traces of the sample, and the memory effects from one sample to the next are minimal. The spectra produced in this method are reproducible in terms of both appearance and sensitivity,[28] and the reduced level of matrix gives rise to a reduced level of background ions. The turbulence at the probe tip is also thought to reduce significantly the suppression effects associated with continuous flow FA/IB.[33] This method can lead to a fast sample throughput, and is suitable for quantification purposes.

The third approach to using continuous flow FA/IB involves on-line separation of the components of the sample *i.e.* on-line HPLC–FA/IB or on-

[30] A. E. Ashcroft, J. R. Chapman and J. S. Cottrell, *J. Chromatogr.*, 1987, **394**, (1), 15.
[31] R. M. Caprioli, *Anal. Chem.*, 1990, **62**, (8), 477A.
[32] 'Continuous Flow Fast Atom Bombardment Data Compilation', Kratos Analytical, Manchester, UK, 1987.
[33] R. M. Caprioli, W. T. Moore and T. Fan, *Rapid Commun. Mass Spectrom.*, 1987, **1**, 15.

line capillary zone electrophoresis – (CZE)–FA/IB, and these two topics are dealt with in the following two Sections.

4 Separation Methods Coupled to Continuous Flow Fast Atom/Ion Bombardment Ionization

Liquid Chromatography

HPLC Solvent Selection

Continuous flow FA/IB is ideally suited for use as an on-line HPLC–MS interface because it is based on the continual flow of solvent into the ionization source. Both isocratic and gradient elutions can be carried out. The flow rate required for continuous flow FAB is typically 1 to 20 μL min^{-1} and so if the output of mobile phase from the column does not fall in this region, it must be adjusted by either flow splitting, or adding a make-up solvent. Many different configurations have been reported for many different columns and flow rates, and Figure 4.8 shows a general schematic summary of coupling continuous flow FA/IB to HPLC.

Both reverse phase and normal phase solvents are suitable, *e.g.* water, acetonitrile, methanol, toluene, and mixtures thereof, and additives such as low percentages (*ca.* 1%) of volatile acids (*e.g.* acetic acid, formic acid, trifluoroacetic acid), bases, salts, and buffers are tolerated. Tetrahydrofuran, dimethyl sulfoxide, and dimethylformamide are not recommended, nor are inorganic acids and bases, and other involatile salts. For a detailed list of solvents and additives which are compatible with mass spectrometric techniques in general, see Section 4 of Chapter 2.

As it is necessary to have a matrix present at the probe tip, it is often convenient to introduce it with the solvent. A matrix such as glycerol will have approximately the same eluotropic strength as methanol, and so the mobile phase can be adjusted accordingly to compensate for this. For instance, if a separation requires 50:50 (*v/v*), water–methanol, then 50:45:5 (*v/v/v*), water–methanol–glycerol should produce similar results. Too much matrix will cause a significant increase in the backing pressure of the HPLC pump, and may also cause some chromatographic peak tailing. If it is not possible to have the matrix in the mobile phase, then it can be introduced by post-column addition. This has been achieved by use of a T-junction, through which the fused silica capillary passes. The matrix, in solution, is added through the third arm of the T-junction, and pumped coaxially along a larger diameter length of silica capillary which also terminates at the probe tip, where mixing of the two mobile phases can take place.[34]

[34] S. Pleasance, P. Thibault, M. Mosely, L. Detering, K. Tomer and J. Jorgenson, *J. Am. Chem. Soc. Mass Spectrom.*, 1990, **1**, 312.

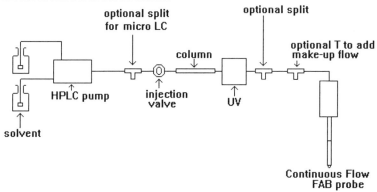

Figure 4.8 *A schematic summary for coupling continuous flow fast atom/ion bombardment ionization mass spectrometry to HPLC*

HPLC Column Selection

The type of column chosen depends on the type of analysis and separation required. Most types of column can be used in conjunction with continuous flow FA/IB, although it should be remembered that the flow rates have to be made compatible, and also if the matrix is present throughout the chromatographic run, then the backing pressure on the HPLC column will be significantly higher. For a comprehensive discussion of HPLC column types see Section 4 of Chapter 2.

Microbore (i.d. 1 mm)[31] and capillary (i.d. 0.3 mm)[27] columns operate optimally at < 50 μL min^{-1} and < 5 μL min^{-1}, respectively, and so are suitable for direct coupling. If gradient elutions are performed with capillary columns, it may well be necessary to form the gradient at a higher flow rate and split down before the mobile phase enters the injector and HPLC column. If such small columns are used, then the complementary couplings and injection valves must also be employed to maintain chromatographic resolution.

For larger i.d. HPLC columns, *e.g.* 2.1 and 4.6 mm, the flow will be too great to pass directly through to the continuous flow FA/IB probe and will have to be split using a suitable splitter. The splitter can be a simple home-made device using a 'T'-piece, or a more elaborate commercially available one (see Section 4 of Chapter 2).

In order to compare off-line chromatography with UV detection directly with a mass spectrometric run, it is often desirable to install a UV detector either in-line *i.e.* in between the outlet of the column and the continuous flow FA/IB probe, or simultaneously, *i.e.* using a splitting device to direct one part of the eluent to the continuous flow FA/IB probe and the other part to the alternative detector. Figure 4.8 shows the former of these two options.

On-line chromatography coupled to continuous flow FA/IB has been used successfully for many applications, including peptides and proteins[30,31,35]

---

[35] R. M. Caprioli, W. T. Moore, B. B. DaGue and M. Martin, *J. Chromatog.*, 1988, **443**, 355.

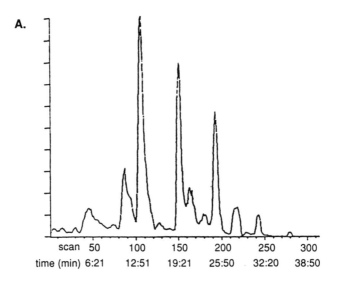

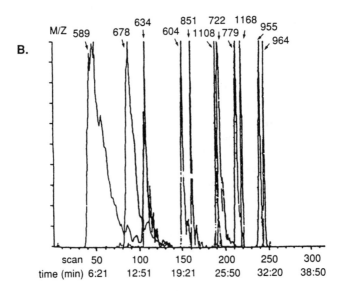

HPLC-FAB MS analysis of the tryptic digest of horse heart cytochrome c. (a) Total ion chromatogram, produced from scans of mass range of 1700-570 dalton. (B) Composite peptide chromatogram, produced from independently normalised selected-ion chromatograms of the [M + H]+ ions of the peptides in the digest.

Figure 4.9 *On-line HPLC–continuous flow FAB for the analysis of peptides* (Reproduced with permission from *J. Chromatogr.*, ref. 35, fig. 3.)

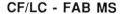

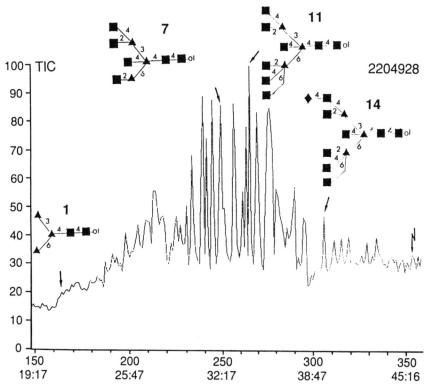

Total ionic current of the permethylated oligosaccharides from hen
ovomucoid on a packed fused-silica capillary column. Oligosaccharides
1, 7, 11, and 14 are localised and their structures are shown.
(■) GlcNAc, (▲) Man, (♦) Gal.

Figure 4.10 *On-line HPLC–continuous flow FAB for the analysis of oligosaccharides*
(Reproduced with permission from *Anal. Biochem.*, ref. 36, fig. 2.)

(Figure 4.9), oligosaccharides[36] (Figure 4.10), and the determination of
acylcarnitines in urine[37] (Figure 4.11), to name just a few.

Finally, continuous flow FA/IB has been coupled to thin layer chromatography (TLC) by the use of a probe which holds a complete TLC plate, pre-
treated with glycerol, inside the ionization source. The TLC plate can then be

[36] P. Boulenguer, Y. Leroy, J. M. Alonso, J. Montreuil, G. Ricart, C. Colbert, D. Duquet, C. Dewaele and B. Fournet, *Anal. Biochem.*, 1988, **168**, 164.
[37] (a) D. L. Norwood, C. A. Bus and D. S. Millington, *J. Chromatogr.*, 1990, **527**, 289; (b) D. S. Millington, D. L. Norwood, N. Kodo, C. R. Roe and F. Inoue, *Anal. Biochem.*, 1989, **180**, 331.

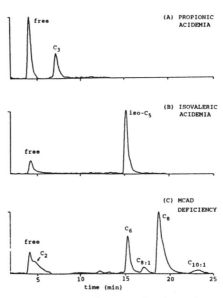

Figure 4.11 *On-line HPLC–continuous flow FAB for the analysis of acyl coenzymes* (Reproduced with permission from *Anal. Biochem.*, ref. 37(b), fig. 7.)

moved so that all the components are passed through the line of the fast atom or ion beam.[38]

Capillary Electrophoresis

Interfacing CE to FAB–MS

Capillary electrophoresis (CE) is a high resolution, very low solvent flow (nL min^{-1}) technique, well suited for coupling with continuous flow FA/IB. An interface is required which not only bridges the gap between the outlet end of the high voltage CE capillary and the inlet of the continuous flow FA/IB probe, but also permits the incorporation of a solvent–matrix solution to enhance the total amount of liquid flowing into the FAB or FIB/LSIMS source up to the required minimum for stable operation, *i.e.* 1 to 2 µL min^{-1}. This has been achieved in two ways: one method utilizes a coaxial probe where the solvent and matrix make-up flow meet with the CE eluent at the probe tip,[39] and the other method involves a liquid junction where the CE eluent and the make-up flow are brought together at a T-junction before entering the probe, and then flow as a mixture contained in one capillary throughout the length of the standard continuous flow FAB probe.[40]

[38] G. W. Somsen, W. Morden and I. D. Wilson, *J. Chromatogr.*, 1995, **703**, 613.
[39] J. S. M. deWit, L. J. Deterding, M. A. Mosely, K. B. Tomer and J. W. Jorgensen, *Rapid Commun. Mass Spectrom.*, 1988, **2**, 100.
[40] R. D. Minard, D. Chin-Fatt, P. Curry and A. G. Ewing, *Proc. 36th ASMS Conf. Mass Spectrom. Allied Topics*, San Francisco, CA, June 5–10, 1988.

Commercial CE–MS interfaces generally involve electrospray ionization (Chapter 2), and if considering setting up a home-made interface such as these for FA/IB operation, particular attention and precautions should be exercised when working with the high voltages (*ca.* 30 kV) required for CE separations.

CHAPTER 5

Field Desorption and Field Ionization

1 What Type of Compounds can be Analysed by Field Desorption and Field Ionization Techniques?

Field Desorption (FD) and Field Ionization (FI) are two closely related techniques, both pioneers during the development of ionization methods appropriate for the analysis of thermally labile and high molecular weight samples. FI and FD emerged as the need grew to develop ionization techniques to complement electron impact, which is limited to relatively low molecular weight samples that can be vaporized prior to ionization and thus often leads to the thermal degradation of involatile or heat sensitive compounds. Since the inception of FI in 1954[1] and FD in 1969[2], the combined techniques reached their height of popularity in the late 1970s to early 1980s before many of their application areas were appropriated by fast atom bombardment mass spectrometry (see Chapter 4), which is generally thought to be a technique that requires less expertize.

FI and FD rely on 'mild' or 'soft' ionization of the sample by an intense electric field applied to the surface on which the sample has been deposited, and have been applied successfully to the following compound classes, traditionally those that can be difficult to analyse by electron impact ionization: polymers,[3] peptides,[4] sugars,[5] organometallics[6] and both organic and inorganic salts.[7] FD is still a method of choice in many polymer applications, although there is some evidence that laser desorption (see Chapter 7) is making progress in this area.

In summary, FI and FD generate predominantly molecular weight information with little fragmentation on samples of low to moderate polarity.

[1] M. G. Ingram and R. Gomer, *J. Chem. Phys.*, 1954, **22**, 1279.
[2] H. D. Beckey, *Principles of Field Ionization and Field Desorption Mass Spectrometry*, Pergamon, London, 1977.
[3] K. Rollins, J. H. Scrivens, M. J. Taylor and H. Major, *Rapid Commun. Mass Spectrom.*, 1990, **4**, 355.
[4] D. M. Desiderio, J. Z. Sabbatini and J. L. Stein, *Adv. Mass Spectrom.*, 1980, **8**, 1198.
[5] J.-C. Prome and G. Puzo, *Org. Mass Spectrom.*, 1977, **12**, 28.
[6] H. R. Schulten, *Int. J. Mass Spectrom. Ion Phys.*, 1979, **32**, 97.
[7] R. P. Latimer and H.-R. Schulten, *Anal. Chem.*, 1989, **61**, 1201A.

2 Field Desorption and Field Ionization

The Principles of Field Desorption and Field Ionization

The nomenclature associated with field ionization and field desorption can appear quite confusing, as the term field ionization is used to describe both an ionization method and an experimental procedure. In practical terms, **field ionization** indicates that the sample has been introduced into the FD/FI source of the mass spectrometer in its vapour state and this can be achieved by using a heated probe, by eluting the sample from a GC–MS interface, or by use of a heated reservoir inlet such as an AGHIS (all-glass heated inlet system). **Field desorption**, on the other hand, refers to the experimental procedure in which a solution of the sample is deposited on the emitter wire situated at the tip of the FD/FI direct insertion probe *prior* to inserting the probe into the ionization source.

By far the most common type of mass spectrometer used for FD/FI analyses is the magnetic sector, and most commercial instruments of this type have FD/FI sources available for purchase; hence the references cited in this Chapter will refer to these mass spectrometers.

The source supplied for FD/FI work will most probably be a combination one suitable for both techniques, and perhaps electron impact or fast atom bombardment as well. The FD/FI source works in conjunction with an **FD/FI insertion probe** which supports a **field emitter**, *i.e.* usually a thin tungsten wire supported on two metal posts that are seated on an insulated, ceramic base. The emitter wire is covered with **microneedles** of pyrolytic carbon or silicon, and is held at the accelerating potential of the ionization source, say 5–10 kV for positive ionization analyses. The probe is inserted into the ionization source through a probe lock, and once positioned in the source (see Figure 5.1) the emitter is within a few millimetres of the **extraction rods** (or an extraction plate, or a counter electrode) which are held at a potential some 12–15 kV lower. Any ions formed, and several possible ionization processes (outlined below) may, or may not, occur in the FD/FI ionization source, can then be extracted into the analyser of the mass spectrometer. The extraction of positive ions in this way is illustrated in Figure 5.2.

When considering the fate of the sample, **field ionization**, which involves the removal of an electron from a molecule by quantum mechanical tunnelling in the presence of a strong electric field, is just one ionization process that is possible. As a result of the voltages applied to the emitter and the extraction plate, the field strengths are greatest near the tips of the microneedles where the radii of curvature are the most pronounced, and this is where field ionization, which requires field strengths of $10^7 - 10^8$ V cm^{-1}, takes place. In the positive ionization mode of operation, FI leads to the production of **molecular ions M$^{+\bullet}$** which have the same nominal mass as the original molecules, from which they differ only by the loss of one electron. In negative ionization mode, molecular anions M$^{-\bullet}$ can be generated by electron capture from the negatively charged needle tip. Most non-polar samples will be ionized by field ionization.

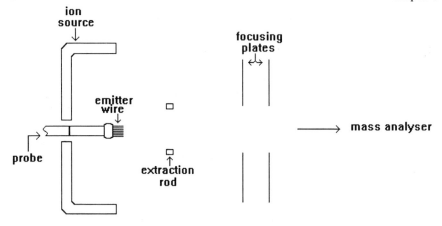

Figure 5.1 *Schematic diagram of a field desorption/field ionization source*

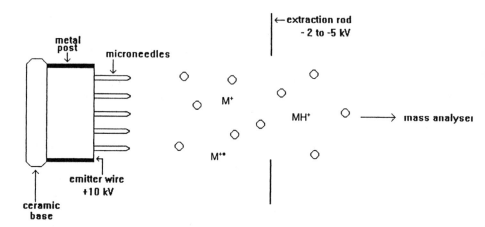

Figure 5.2 *Field desorption/field ionization ion formation and extraction*

Cation or **anion attachment, thermal emission,** and **proton abstraction** are the other ionization processes that can take place in the FD/FI source.

Basic, polar organic molecules (*e.g.* amines) are most likely to be analysed in positive ion mode where they can form **cation attachment** ions in the condensed phase with cations such as H^+ and Na^+. The combination of emitter heating and the strong field then results in the desorption of these MH^+ and MNa^+ cations, which are subsequently extracted and analysed. In negative ion mode, it is possible for some polar, acidic samples (*e.g.* carboxylic acids) to ionize by **anion attachment,** for example with anions such as chloride, Cl^-, which produces $(M + Cl)^-$ ions.

Both organic and inorganic salts are species containing pre-formed ions, and the most probable outcome for these samples when they are applied directly to the emitter is **thermal emission,** with subsequent analysis of the cations or anions present depending on whether positive or negative ionization respectively is used. For example, a salt of the general formula C^+A^- will generate C^+ ions in positive ion mode, and A^- ions in negative ion mode.

Proton abstraction is another commonly encountered mode of ion formation in negative ionization, and most polar organic molecules susceptible to negative ionization readily produce $(M-H)^-$ ions. The four possible ionization modes and the type of sample generally associated with each are summarized in Table 5.1.

Practical Operation of Field Desorption and Field Ionization

Essential Requirements for Operation

The following is a list of necessary equipment for FD/FI operation:

- magnetic sector mass spectrometer under good vacuum, equipped with an FD/FI ionization source, and any other equipment that may be required for the introduction of volatile samples into the ionization source, *e.g.* a reservoir inlet or a gas chromatograph;
- an FD/FI probe with a supply of emitters, usually carbon or silicon; emitters are commercially available, or can be made with suitable preparation apparatus (see Chapter 5, Section 2);
- a range of volatile solvents to dissolve the samples, *e.g.* acetone, dichloromethane, toluene, methanol/water mixtures;
- a syringe to load the emitter with the sample;
- a suitable reference material for optimization *e.g.* acetone, tris(perfluoromethyl)-*s*-triazine.

Setting up Field Desorption and Field Ionization

The **emitter,** or more specifically the **emitter needles,** play an important rôle in **FD/FI** mass spectrometry, and great care must be excercised in its preparation if it is to be constructed successfully in the laboratory. It is recommended that extensive further reading regarding the preparation of emitters be undertaken before any experimentation takes place.[2,8]

Carbon emitter needles are the most frequently encountered type and are prepared on a thin tungsten wire (10 μm diameter) using benzonitrile in a specially set-up vacuum rig. A good number of microneedles are required on the wire in order to provide many ionization sites and hence produce high ion yields. By maintaining the wire at a high temperature (1200 °C) in benzonitrile vapour, decomposition of the nitrile onto the hot wire produces fine carbon growths having small radii of curvature. When an electric current is applied to this wire, these growths become substantially thicker and branch out. After several hours the carbon growths are approximately 30 μm in length, and

---

[8] W. D. Lehmann and R. Fischer, *Anal. Chem.*, 1981, **53**, 743.

Table 5.1 *Ionization methods encountered in field desorption and field ionization techniques*

Ionization mechanism	Ions formed	Ionization mode	Type of sample
Field ionization, FI	$M^{+\bullet}$	+	non-polar organics
	$M^{-\bullet}$	−	
Cation attachment	MH^+, MNa^+	+	basic, polar organics
Anion attachment	$(M + Cl)^-$	−	acidic, polar organics
Thermal emission	C^+	+	ionic species
	A^-	−	*e.g.* salts, C^+A^-
Proton abstraction	$(M - H)^-$	−	acidic, polar organics

although the emitter wires are fragile, with careful treatment they can last for some time.

Emitters with shorter needles can be prepared more easily simply by heating an unprepared emitter wire (in position on a FD/FI probe) in the ionization source for several seconds.[9] The main disadvantage of this type of emitter wire is the decreased surface area, which increases the difficulty of loading the solute.

Before setting up for FD/FI analyses, the source temperature should be set at 60 °C or 180 °C for FD or FI respectively. Optimization for FD/FI analyses can then be achieved by monitoring and tuning the ion beam arising from a known sample or solvent with an oscilloscope or on the data system display. Such a sample can be introduced through a reservoir inlet system such as an All Glass Heated Inlet System (AGHIS) or a reference inlet to produce a steady ion beam for a long enough period of time to allow optimization of the ionization conditions. The AGHIS heated reservoir inlet system allows a continual flow of a volatile sample to enter the source in its vapour form, and is particularly popular in the petroleum industry. The solvent acetone, for example, is useful for tuning purposes, and it produces $M^{+\bullet}$ ions at *m/z* 58 in positive ionization operation without any heat being applied to the emitter. The source accelerating voltage is usually set according to the mass spectrometer being used after taking into account the molecular weights of the samples which are to be analysed; if high molecular weight samples are under investigation, the source accelerating voltage may have to be lowered. The extraction rod potential is then set so that it is some 10–15 kV lower than the accelerating voltage. Other source tuning parameters can now be adjusted until optimum performance is achieved, and then the flow of acetone should be stopped. At this stage it is pertinent to heat the emitter a little in case any acetone has lodged on the emitter needles, which would result in contamination of subsequent samples.

For the **field desorption** analysis of samples, the **sample** is dissolved in an appropriate, volatile solvent such as acetone, dichloromethane, or toluene to a concentration of *ca.* 1 mg mL^{-1}. If the sample is only soluble in water, it

[9] U. Giessmann, H. J. Heinen and F. W. Röllgen, *Org. Mass Spectrom.*, 1979, **14**, 177.

should first be dissolved and then diluted with a little of a miscible organic solvent such as methanol to decrease the surface tension of the solvent. If it is impossible to dissolve the sample, then a suspension can often produce reasonable results. The sample and solvent are then applied to the emitter wire, either by dipping the wire in the solution or by applying the solution to the wire with a syringe. If the wire is dipped into the solution, contamination of the posts and ceramic base can occur, and may cause memory effects with subsequent samples. Dipping also uses rather more sample than direct application.

With the sample applied to the emitter wire, the FD/FI probe can then be inserted into the ionization source of the mass spectrometer, with careful use of the probe lock and the affiliated rough pumping line. During this procedure, most of the solvent will be evaporated, leaving the sample alone on the emitter wire. The emitter is heated to optimize ionization of the sample, or the current can be programmed (*e.g.* from 0 to 60 mA) to increase with time for the analysis. The extraction rods can also be heated to avoid excessive contamination and a typical temperature is *ca.* 150 °C. Once the sample is in place, a data acquisition is initiated simultaneously with the temperature programming of the emitter. It is usually necessary to collect and average several scans as the ion current can be quite weak with this type of analysis. Often with an unknown sample, the emitter is heated until sample desorption begins to take place and a stable ion beam is achieved, and then it is held at this temperature. Further heating may well increase the ion beam intensity, but can lead to thermal fragmentation of the molecule.

For **field ionization** analyses involving reasonably volatile samples, the compound(s) under investigation is volatilized by heat close to the emitter so that its vapour can condense onto the emitter needles, and hence FI is suitable for on-line gas chromatography–mass spectrometry whereby the GC effluent containing the separated samples in a flow of heated helium gas is taken directly into the ionization source where it can pass over the emitter wire. Alternatively, the use of a reservoir inlet will exude vaporized samples into the ionization source.

The Analysis of Samples

Fragmentation is usually limited in FD/FI analyses as there is little excess energy from these soft ionization processes, and predominantly molecular mass information is generated. Mixtures of compounds can be screened by applying the total mix to the emitter and interpreting the resulting spectrum which, under appropriate conditions, will contain one diagnostic ion for each component. The precise spectrum obtained from any sample will depend on the ionization processes occurring, and also the method of introducing the sample into the ionization source.

The differences between field desorption and field ionization can be illustrated with reference to their respective spectra obtained for a sample of D-glucose, molecular weight 180 da (see Figure 5.3). Figure 5.3(b) shows the field desorption spectrum produced by coating the sample onto the emitter

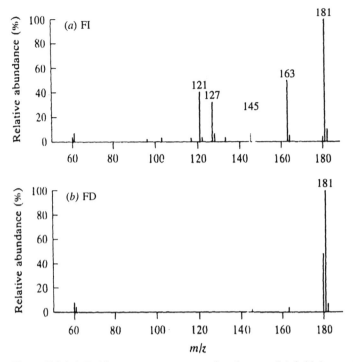

Figure 5.3 *(a) Field ionization spectrum of D-glucose; (b) field desorption spectrum of D-glucose*
(Reproduced with permission from Micromass UK Ltd.)

outside the mass spectrometer followed by insertion of the probe into the ionization source and then application of the ionizing voltage. The spectrum is dominated by MH$^+$ ions at *m/z* 181, produced by cation attachment of a proton, H$^+$, to the polar saccharide compound, accompanied by lower intensity M$^{+\bullet}$ ions at *m/z* 180. The same sample analysed under field ionization conditions produces not only the MH$^+$ ions, but also fragment ions of significant intensity at *m/z* 163, 127, and 121 [see Figure 5.3(a)]. Not only does the sample undergo protonation, but also the heat involved in volatilizing the sample in order to introduce it into the ionization source via the GC causes fragmentation.

If a sample is prone to cationization,[10] then it may be difficult to interpret the spectrum and conclude whether MH$^+$ or MNa$^+$ or MK$^+$ ions have been formed. In these cases it may be necessary to intentionally add another cationic species to the sample whilst in solution. Some samples, for example certain sugars, may produce weak spectra unless cations in the form of an acid (*e.g.*

[10] J. R. Chapman, in *Practical Organic Mass Spectrometry*, John Wiley & Sons, Chichester, UK, 1985.

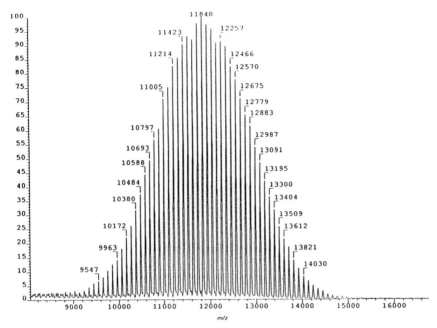

FDMS profile of polystyrene sample D.

Figure 5.4 *Field desorption spectrum of a polystyrene of average molecular weight 12 000 da*
(Reproduced with permission from *Rapid Commun. Mass Spectrom.*, ref. 3, fig 5)

dilute hydrochloric acid), or sodium salts such as the chloride or iodide are added to the sample in solution before applying it to the emitter.

One of the main uses of FI is for the analysis of hydrocarbons, which are commonly admitted to the ionization source via a reservoir inlet system. Alkyl and aryl hydrocarbons of the general formula C_nH_{2n+z}, where n is an integer value and z can range from large negative integer values to +2, depending on the degree of unsaturation in the molecule, are all suitable for analysis. These molecules generate $M^{+\bullet}$ ions with little fragmentation, in contrast to their electron impact spectra which generally contain weak molecular ions accompanied by substantial fragmentation. FI has a further advantage over electron impact in this application field in that it can be used for hydrocarbon samples of much lower volatility and hence higher molecular weight. Usually all the components of a mixture can be detected because the suppression effects associated with FI are minimal.

More intractable organic polymers are better suited to FD which avoids the need for volatilization of the sample. The molecular weight distribution of oligomers can be determined from their spectra, as can the end groups of the polymer chains. To illustrate the molecular weight determinations achievable with FD, Figure 5.4 shows the mass spectrum of a polystyrene sample of

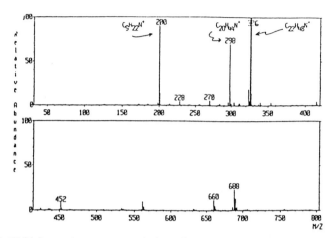

Figure 5.5 *Field desorption spectrum of a biocide containing tetraalkylammonium salts*
(Reproduced with permission from *Anal. Chem.*, ref. 11, fig 1)

average molecular weight 12 000 da which was applied to the emitter as a
solution in toluene.[3] Using an emitter current increasing from 0 mA to 40 mA
with an ionization source accelerating voltage of 7 kV and an extraction
voltage of −5 kV, the mass spectrometer was scanned at 50 seconds mass-
decade[-1] over the range m/z 8 000 to 17 000 to produce this spectrum. As can
be seen, the molecular weight distribution and the mass separation between
oligomer ions can be determined readily from the molecular ion species
present. The number-average molecular weight (M_n) ascertained from the
FD–MS method was 11 900 da, which was consistent with the gel permeation
chromatography finding of 12 100 da. The main limitation of polymeric
analyses such as this one is the solubility of the often intractable samples, and
frequently the analyst has to resort to hot dimethyl sulfoxide – hardly a solvent
of choice!

The FD analysis of quaternary ammonium salts is illustrated with an
example of a biocide containing three major components, the tetraalkylammonium chloride salts $C_9H_{22}N^+Cl^-$, $C_{20}H_{44}N^+Cl^-$, and $C_{22}H_{48}N^+Cl^-$, with
cationic molecular weights of 200, 298, and 326 da respectively.[11] Biocides are
commonly used in oil field operations. The FD spectrum in Figure 5.5 shows
M^+ ions for all three components, produced by thermal emission of the
already-charged species. The sample was applied in solution to the emitter, and
during the analysis the emitter temperature was set with a constant 20 mA
current.

FD/FI has gained a reputation of being difficult to practise, the main
problems being associated with the emitter, which requires skill to make or
which can be expensive to purchase. The quality of the emitter is of great
importance, especially as ion currents tend to be lower than with other

[11] J. Shen and A. S. Al-Saeed, *Anal. Chem.*, 1990, **62**, 116.

ionization techniques. However FD/FI techniques are still in worthy operation in such areas as polymer and petroleum analyses. Other types of sample analyses that were once popular, such as peptide analyses, are now more routinely accomplished by the use of fast atom bombardment, or more recently, electrospray ionization.

CHAPTER 6

Thermospray Ionization

1 What Type of Compounds can be Analysed by Thermospray Ionization?

The years preceding the advent of electrospray were the real heyday for thermospray (TSP): when it had been developed fully, and reliable, uninterrupted analyses could be performed routinely. Although its popularity has waned somewhat in preference for newer, more robust API techniques, many laboratories continue to use TSP to produce some very useful data. The combined tolerance of TSP and closely related techniques to both polar and non-polar solvents means that when coupled to HPLC, both normal and reverse-phase separations can be carried out with isocratic and gradient elutions.

Thermospray is a soft ionization technique producing predominantly MH^+ or $(M-H)^-$ ions, which may be accompanied by some degree of fragmentation under certain conditions. Thermospray is best suited to the analysis of organic compounds with molecular masses of less than 1000 da that exhibit some polarity. Into this wide category fall molecules such as drugs and drug metabolites,[1] on which some rather elegant studies have been carried out, pesticides,[2] bile acids,[3] environmental pollutants,[4] peptides,[5] lipids,[6] and alkaloids.[7]

2 Thermospray Ionization

The Principles of Thermospray Ionization

Thermospray ionization involves the introduction of a relatively high flow $(0.2-2 \text{ mL min}^{-1})$ of solvent into the ion source of a mass spectrometer and so

[1a] I. G. Beattie and T. J. A. Blake, *J. Chromatogr.*, 1989, **474**, 123.
[1b] T. J. A. Blake and I. G. Beattie, *Biomed. Environ. Mass Spectrom.*, 1989, **18**, 637.
[1c] T. J. A. Blake and I. G. Beattie, *Biomed. Environ. Mass Spectrom.*, 1989, **18**, 860.
[1d] T. J. A. Blake and I. G. Beattie, *Biomed. Environ. Mass Spectrom.*, 1989, **18**, 872.
[1e] A.E. Ashcroft, *Anal. Proc. (London)*, 1991, **28**, (6), 179.
[2] D. Barcelo, *Biomed. Environ. Mass Spectrom.*, 1988, **17**, 363.
[3] C. Eckers, N. J. Haskins and T. Large, *Biomed. Environ. Mass Spectrom.*, 1989, **18**, 702.
[4] T. A. Bellar and W. L. Budde, *Anal. Chem.*, 1988, **60**, 2076.
[5] R. D. Voyksner and T. W. Pack, *Biomed. Environ. Mass Spectrom.*, 1989, **18**, 897.
[6] N. W. Rawle, R. G. Willis and J. D. Baty, *Analyst (London)*, 1990, **115**, 521.
[7] S. Auriola, T. Naaranlahti, R. Kostiainen and S. P. Lapinjoki, *Biomed. Environ. Mass Spectrom.*, 1990, **19**, 400.

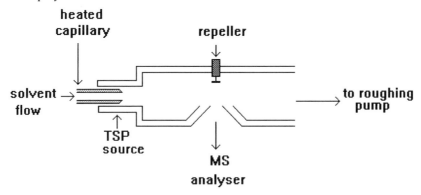

Figure 6.1 *Typical thermospray ionization source and interface*

is ideal as an interface for HPLC–MS using standard bore columns. The legend behind TSP is that during experiments in which a liquid flow was introduced into a modified chemical ionization source, ions were still observed even when the filament was (unknowingly) switched off![8] A typical TSP source complete with the probe interface inserted is shown in Figure 6.1. The probe acts as a transfer line for taking the solvent and solute into the source. As the source is held under vacuum, the amount of solvent entering the source must be reduced significantly in order to maintain this vacuum, and so an extra roughing vacuum pump line is fitted to the source, usually directly opposite the probe, for this purpose.

The solvent is pumped down the resistively **heated capillary tube** or **vaporizer** (usually 100 μm internal diameter) inside the probe interface, the tip of which is situated in the source. The source itself is heated to prevent condensation of the solvent, and the temperature of the capillary is adjusted until the solvent is vaporized. This generates a jet of vapour which contains small, electrically charged droplets if the solvent is at least partially aqueous and contains an **electrolyte** (see Figure 6.2). The presence of the electrolyte is particularly important, as without this the ion yields, and hence sensitivity, are quite low. For this purpose, a volatile salt is used and more often than not 50–100 mM aqueous ammonium acetate is employed. The droplets continue to vaporize and, due to the heat and vacuum, become smaller in size as they travel through the source. Eventually they are sufficiently small for free ions to be expelled from the surface of the droplets. Positive and negative ions are generated depending on the gas phase acidity or basicity of the solvent and solute. These ions leave the source through a **sampling cone**, which has an orifice in the centre, and this process is encouraged by an **ion repeller** that is situated directly opposite the sampling cone. The ions are then analysed by the **analyser** of the mass spectrometer, which is generally either a quadrupole or a magnetic sector

[8] C. R. Blakey and M. L. Vestal, *Anal. Chem.*, 1983, **55**, 750.

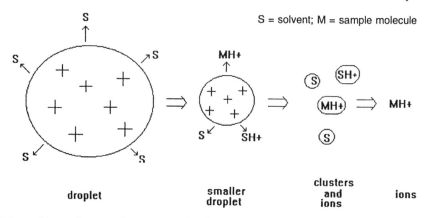

Figure 6.2 *Production of positive ions by thermospray ionization*

instrument. A high voltage applied to the repeller will tend to induce fragmentation of the sample ions, and these fragment ions will also be analysed by the mass spectrometer.

If an electrolyte cannot be used for a particular analysis, for example if a normal-phase separation is being carried out using organic solvents in which ammonium acetate is insoluble, then an electrical discharge in the source may be employed as an alternative way to effect ionization. Most thermospray sources are fitted with such a discharge electrode, and when this is used, the technique becomes formally known as **filament-on operation** or **plasmaspray (PSP)** ionization (see Figure 6.3). Filament-on ionization is a gas phase ionization technique, in which the solvent vapour is first ionized by the electrical discharge (usually in the region of 500–1000 V) to form a reagent gas plasma, similar to that produced during chemical ionization (see Chapter 3). The solute molecules are in turn ionized by the solvent ions. Positive ions are generally formed by protonation and negative ions by deprotonation, and so the relative proton affinities of the solute and the solvent determine the efficiency of the ionization process. Another feature of filament-on operation is that some degree of fragmentation takes place, and these fragment ions are also analysed by the mass spectrometer.

In reality, many analyses are a mixture of TSP and filament-on TSP, especially those employing on-line, reverse-phase, HPLC–MS using gradient elutions where the first part of the run is officially termed TSP, and the latter part, when the aqueous content of the eluent has decreased to such an extent that the ionization process is failing and the discharge electrode needs to be switched on, filament-on TSP. Hence this Chapter will deal with both TSP and filament-on TSP together, as their usage is so intertwined that it is difficult to separate them.

Solvents used successfully in TSP operation include water, methanol, acetonitrile, propan-2-ol, dichloromethane, and hexane, to name just the most popular ones. Tetrahydrofuran has been reported to cause problems including

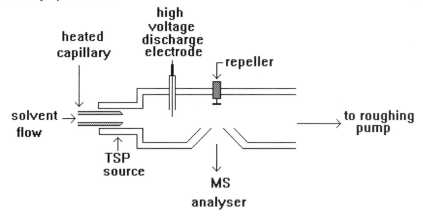

Figure 6.3 *Typical filament-on thermospray ionization source and interface*

blockage of the capillary.[9] As already stated, ammonium acetate is the most regularly used electrolyte or buffer, but in reality any *volatile* buffer may be employed including ammonium formate, ammonium hydroxide, acetic acid, triethylamine, and diethylamine. Involatile buffers and inorganic acids should be avoided at all costs, and this list includes phosphoric acid, phosphate salts, mineral acids, perchloric acid, and perchlorates. Not only will these result in blocked capillaries and dirty sampling cones, some are downright dangerous, especially perchlorates, which have caused severe explosions after precipitating on the heated source chamber.

Practical Operation of Thermospray Ionization

Essential Requirements for Operation

Thermospray ionization sources tend to be supplied with quadrupole or magnetic sector mass spectrometers, and this includes hybrid or tandem combinations of these two analysers. Most thermospray sources have a discharge pin permanently assembled for filament-on operation, so that once the source is fitted in the mass spectrometer either technique can be carried out with equal ease and there is no further need for source changing. The following is a list of essentials for thermospray operation:

- a working quadrupole or magnetic sector mass spectrometer, fitted with a clean thermospray source, which may or may not have a discharge electrode. There will probably be an extra roughing vacuum pump for the source, and both the source region and the analyser should be under vacuum;

[9] D. A. Catlow, *J. Chromatogr.*, 1985, **323**, 163.

- an HPLC solvent delivery pump, capable of producing a pulse-free flow of 1 mL min^{-1} in either isocratic or gradient mode;
- an injection valve, preferably with several injection loops, *e.g.* a 20 μL loop for sample injections and a 100 μL loop for tuning and calibration purposes;
- a range of HPLC syringes and fittings, including an in-line filter;
- a range of HPLC grade solvents suitable for the analyses in mind;
- ammonium acetate or another appropriate, volatile electrolyte;
- a sample suitable for tuning, *e.g.* adenosine for TSP, or caffeine for filament-on TSP;
- a suitable calibrant, *e.g.* a polyethylene glycol.

Setting up and Using Thermospray

In order to check out thermospray operation, a reverse-phase solvent system should be mixed, sparged and set to flow (preferably via a filter) through the thermospray probe on the bench at a rate of 1 mL min^{-1}. A 50:50 (*v/v*) water–methanol or acetonitrile mix containing 0.1 M ammonium acetate is ideal for this purpose. Initially the system should be set up *without* an HPLC column in-line so that the source can be tuned up quickly and the mass spectrometer calibrated, if necessary. Fault finding is also much simpler without an HPLC column in place. Once a flow of solvent throughout the HPLC system and the thermospray probe has been established, and no leaks or blockages detected, then the flow can be temporarily stopped and the probe inserted through the vacuum lock into the source of the mass spectrometer. If solvent is flowing during this operation, an excessive rise in the pressure in the source will be observed. When removing the probe, the solvent flow should first be stopped and then, after a period of 2 or 3 min to allow all residual solvent to evaporate, the probe can be removed through the direct insertion lock.

Before switching the solvent flow on again, the **temperature** of the TSP source should be checked. It needs to be high enough to prevent condensation of solvent on the source, and is set typically in the region 200–300 °C. Too high a temperature can lead to the thermal degradation of labile compounds, and so should be avoided whenever possible. Check also that the extra source rough vacuum pump is running on gas ballast. The capillary interface, or vaporizer temperature should be set initially to *ca.* 200 °C. During analyses, the source temperature is usually kept constant, while the capillary temperature is varied according to the sample under scrutiny, and the composition and flow rate of the mobile phase. Now the solvent flow can be initiated, and it is customary to step the flow, starting from say, 0.2 mL min^{-1}, to 0.5 mL min^{-1}, and finally to 1 mL min^{-1}. As the solvent vapour reaches the ionization source, the pressure in the source will rise, and this can be observed on the source vacuum gauge. If the pressure is pulsing, then the capillary (or vaporizer) temperature should be increased until the pressure remains steady.

When the source pressure has settled down, the source voltages can be switched on, and a number of background ions should be apparent immedi-

Table 6.1 *Commonly observed TSP positive ionization background ions from a mobile phase of 1:1 (v/v) acetonitrile–water with 0.1 M ammonium acetate added*

Observed ion m/z	Inference
18	NH_4^+
59	$CH_3CN.NH_4^+$
78	$(CH_3CO_2NH_4)H^+$
100	$(CH_3CN)_2.NH_4^+$
141	$(CH_3CN)_3.NH_4^+$

Table 6.2 *Commonly observed TSP negative ionization background ions from a mobile phase of 1:1 (v/v) acetonitrile–water with 0.1 M ammonium acetate added*

Observed ion m/z	Inference
59	$CH_3CO_2^-$
119	$(CH_3CO_2H.CH_3CO_2)^-$
137	$(CH_3CO_2H.H_2O.CH_3CO_2)^-$
160	$(CH_3CO_2H.CH_3CN.CH_3CO_2)^-$
179	$[(CH_3CO_2H)_2.CH_3CO_2]^-$

ately. Solvent related ions are commonplace, and the most frequently observed ones for both positive and negative TSP ionization are listed in Tables 6.1 and 6.2 respectively. The ammonium acetate MH^+ ions at *m/z* 78 should be visible, and can be monitored and optimized. This is a simple method of optimizing TSP operation initially, and although it will provide a quick answer to the stability and operability of the system, it is always advisable to inject a standard sample to use for further tuning purposes prior to analysing unknown samples. Adenosine and caffeine are suitable compounds to tune on for TSP and filament-on TSP respectively. Standard solutions of both can be made by dissolving 1–10 µg mL^{-1} in water. These solutions can be injected repeatedly through an appropriate injection loop, or can be pumped continuously into the ionization source as deemed necessary. Both produce MH^+ ions in positive ionization mode, adenosine at *m/z* 268 and caffeine at *m/z* 195.

When tuning for TSP operation with adenosine, the **discharge electrode** or **filament** should be switched off. The solvent should be water or miscible mixtures of aqueous organic solvents, and ammonium acetate (50–100 mM) should be present in the mobile phase. While monitoring the *m/z* 268 ions of adenosine (Figure 6.4), the position of the probe within the ionization source should be checked. Peak height and stability are both of importance when tuning for TSP operation.

Now the **capillary** or **vaporizer** temperature can be checked; this is generally in the range 150–250 °C. The vaporizer temperature is one of the most critical

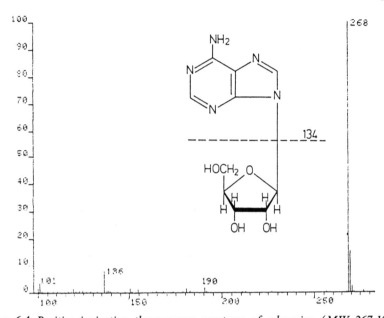

Figure 6.4 *Positive ionization thermospray spectrum of adenosine (MW 267.10 da)*
introduced in 0.1M aqueous ammonium acetate solution; spray temperature
225 °C,
(Reproduced with permission from Elsevier, J. R. Chapman, *J. Chromatogr.*,
1985, **323**, 153, figure 5)

parameters in TSP operation and should be optimized for every different type
of sample, wherever possible. Too low a vaporizer temperature will result in
the solute and solvent spraying into the ionization source in their liquid form,
and hence gas phase ions will not be formed. Too high a vaporizer temperature
causes premature evaporation of the solute and solvent before they reach the
outlet of the capillary and this results in an unstable, pulsing ion beam. As ion
formation in TSP operation depends very precisely on the extent of desolvation
and the energy of the nebulized droplets, the outcome of this is that an
inappropriate vaporizer temperature will lead to a loss of sensitivity. For an
unknown sample, the vaporizer temperature should be set to a value at the
lower end of the range outlined above, and then gradually increased until
sample ions start to appear. An optimum temperature will be reached, and it is
important not to exceed this temperature. It is more practical to have a stable
ion beam with, say, 95% of the optimum sensitivity then an unstable beam
with (occasionally) 100% optimum performance.

The **repeller** also exhibits a significant impact on the sensitivity of TSP
operation. In general, the higher the repeller voltage, the more efficient the
extraction of ions from the ionization source, and hence the higher the
sensitivity. At the higher repeller voltages however, fragmentation is induced
and these fragment ions will be analysed together with the quasimolecular
ions. This can lead to the production of structurally informative ions in the

mass spectrum and is useful for qualitative studies. If quantitative studies are being carried out then careful optimization is required to minimize any source-induced fragmentation.

Any other source tuning parameters particular to the specific type of mass spectrometer in use should be optimized.

In general, filament-on operation is used when the aqueous content of the mobile phase is less than 20%, or if an electrolyte is to be avoided for either chromatographic purposes or for the simple fact that it will not dissolve in the mobile phase if non-aqueous solvents are being used. If filament-on TSP ionization is to be used, then a further tuning parameter needs to be taken into account, the **discharge electrode**, or **filament**. Caffeine (MW 194.08 da) is a useful standard to tune on, especially as the mobile phase will not necessarily contain an electrolyte and so ammonium acetate ions may not be present. Again a standard solution can be introduced either by repetitive injections or by continuous infusion. The discharge or filament should be switched on and slowly increased until the optimum value has been determined, which is often in the region of 500–1000 V. Then all of the tuning parameters outlined above should be re-checked. The vaporizer temperature may need to be increased slightly to ensure that all of the solute and solvent molecules are in the gaseous state. Sometimes a slightly lower flow rate than that required for TSP gives rise to better performance in filament-on operation. The harsher nature of filament-on TSP ionization often leads to the induction of fragment ions, as can be seen by comparing the two spectra of caffeine,[10] one obtained under TSP conditions and the other with filament-on ionization (see Figure 6.5). The former spectrum clearly shows MH^+ ions at *m/z* 195, while the latter exhibits ions consistent with fragmentation of the molecule, such as those at *m/z* 137 which represent loss of 58, or C_2H_4NO, from the intact molecule.

Another reported cause of fragmentation in thermospray operation is an increase in the concentration of the electrolyte.[11]

The Analysis of Samples

Once the system, be it thermospray or filament-on TSP, has been optimized carefully and the ion beam is stable, then 'real' samples can be analysed. The type of analysis will to some extent determine the conditions used: whether it is a quantitative or qualitative study, whether molecular weight and/or fragmentation information is required, whether aqueous or organic solvents are appropriate for dissolving the sample, and any on-line chromatography that may be carried out.

However, there are three methods for introducing samples into the ionization source: direct infusion, loop injection, or loop injection followed by on-line chromatography (Figure 6.6). Loop injection represents a rapid method

[10] J. R. Chapman and J. A. E. Pratt, *J. Chromatogr.*, 1987, **394**, 231.
[11] G. Schmelzeisen-Redeker, M. A. McDowall, U. Giessmann, K. Levsen and F. W. Röllgen, *J. Chromatogr.*, 1985, **323**, 127.

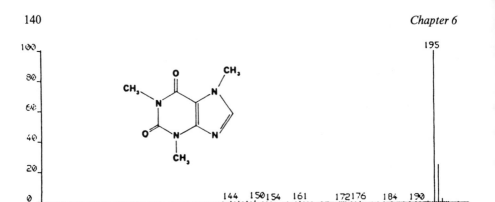

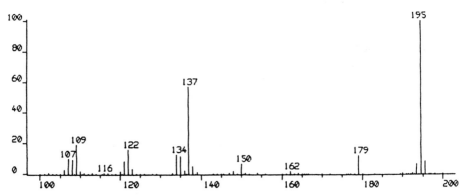

Figure 6.5 *Positive ionization spectra of caffeine (MW 194.08 da) obtained by (a) thermospray; and (b) filament-on thermospray ionization.*
(Reproduced with permission from Elsevier, ref. 10, fig. 6)

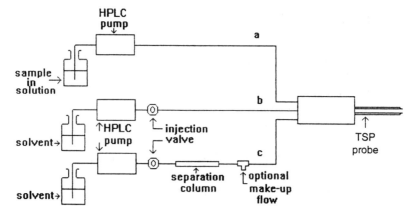

Figure 6.6 *Summary of methods of sample introduction into a thermospray ionization source: (a) continuous sampling; (b) loop injection; (c) loop injection followed by on-line chromatography*

for obtaining spectra from a good number of samples that do not require extensive purification, whereas direct infusion, which requires continual pumping of the sample into the mass spectrometer, is limited to a much smaller number of samples due to the time taken to set up one sample and thoroughly rinse all transfer lines afterwards. Direct infusion of a sample means that the sample is pumped through the solvent delivery pump, and the possibility of contamination of the pump and of subsequent samples must be considered. Some of the perils and pitfalls of coupling separation methods to TSP will be dealt with later (see Section 3 of Chapter 6), whereas this Section will deal with non-chromatographic sample analyses.

For loop injection, the sample should be dissolved in either the same solvent as the mobile phase, or at least one that is totally miscible with it. If the sample drops out of solution, it will almost certainly cause a blockage in the narrow thermospray capillary. A blockage of the capillary can be detected from a complete loss of ion signal on the mass spectrometer, accompanied by a substantial increase in the backing pressure monitored by the HPLC pump, and usually means that the capillary has to be replaced, unless it can be cleared by rinsing with a range of appropriate solvents. This latter operation should be performed with the probe removed from the ionization source to prevent possible source contamination.

Due to the high flow rates employed for thermospray ionization, it will take only several seconds for the sample to be transported from the injection valve to the ionization source, and the signal from the sample will probably last for less than a minute, which is why continuous sample infusion or large volume injection loops are often used for tuning purposes.

If at all possible, a standard which is as chemically close as possible to the samples under investigation should be used to tune up on before acquiring data. For example, if a drug metabolite investigation is about to be performed, then a sample of the parent drug or any other readily available similar metabolite should be used to set the source and capillary parameters.

As with most 'soft' ionization techniques, positive ion operation tends to produce MH^+ ions, while negative ionization generates $(M-H)^-$ ions. However due to the continual presence of ammonium acetate in the majority of TSP analyses, adducting is quite commonplace, with MNH_4^+ and $[M(OCOCH_3)]^-$ ions being the most frequently encountered ones in positive and negative ion modes respectively. Solvent–sample complexes are also possible, and these can complicate the spectrum. For example $[M + CH_3CN]H^+$ or $[M + CH_3CN]NH_4^+$ may well be present if a water–acetonitrile mobile phase is in use with ammonium acetate present as the electrolyte. A list of some of the different types of molecular ion related species that have been observed in positive ionization thermospray operation using either water–acetonitrile or water–methanol mobile phases in the presence of ammonium acetate is presented in Table 6.3.[12] If more than one such species is present, and this is usually the case, then it is possible to predict the molecular mass of

[12] H. Maeder, *Rapid Commun. Mass Spectrom.*, 1990, **4**, 52.

Table 6.3 *Some commonly observed positive TSP ionization molecular ion related species from a mobile phase of either 1:1 (v/v) methanol–water or acetonitrile–water with 0.1 M ammonium acetate added*[12]

Observed ion in relation to M, the molecular mass m/z	Inference
$(M - 18)^+$	$(M + NH_4 - 2H_2O)^+$
$(M - 17)^+$	$(M + H - H_2O)+$
M^+	$(M + NH_4 - H_2O)^+$
$(M + 1)^+$	$(M + H)^+$
$(M + 18)^+$	$(M + NH_4)^+$
$(M + 33)^+$	$(M + H + CH_3OH)^+$
$(M + 42)^+$	$(M + H + CH_3CN)^+$
$(M + 50)^+$	$(M + NH_4 + CH_3OH)^+$
$(M + 59)^+$	$(M + NH_4 + CH_3CN)^+$
$(M + 78)^+$	$(M + H + NH_4OAc)^+$
$(2M + 1)^+$	$(2M + H)^+$
$(2M + 18)^+$	$(2M + NH_4)^+$

the compound under investigation with some certitude. If, however, just one of these species appears in the spectrum, some doubt is cast upon its interpretation.

One drawback of thermospray ionization is that it does tend to exhibit high levels of background ions due to the solvents and electrolyte present, and these should not be confused with sample-related ions. Commonly observed solvent-related background ions in both positive and negative ionization modes from a standard 1:1 (v/v) acetonitrile–water mobile phase with 0.1 M ammonium acetate added can be found in Tables 6.1 and 6.2 respectively.[2] The extent of these ions can limit the lower end of an acquisition to m/z 150–200; below this, sample ions are often difficult to detect and so samples with molecular masses of below 200 da can go undetected.

Samples with molecular masses of more than 1000 da can present difficulties too,[13] and although more recent source modifications to commercial instruments have alleviated this problem, it is still unusual to see spectra reported on real samples, as opposed to cluster ions and polymeric materials, above 1500 da.

Calibration in positive ionization TSP using a combined solution of polyethylene glycol 600 (0.1–1 mg mL^{-1}) and ammonium acetate (7.7 mg mL^{-1}) in water affords predominantly a series of ammoniated ions of the general formula $H(CH_2CH_2O)_nOH.NH_4^+$, where adjacent members of the series differ by 44 da, which covers the range up to *ca.* m/z 800 (see Appendix 2).

Accurate mass measurements have been carried out on a magnetic sector mass spectrometer using polyethylene glycol as an internal reference material[3,14] introduced concomitantly with the sample, and errors within 5 mmu

[13] R. H. Robbins and F. W. Crow, *Rapid Commun. Mass Spectrom.*, 1988, **2**, 30.
[14] L. Baczynskyj and G. E. Bronson, *Biomed. Environ. Mass Spectrom.*, 1988, **16**, 253.

have been reported. In this example, the reference material was added to the mobile phase after the chromatographic separation of the mixture.

3 Separation Methods Coupled to Thermospray

Liquid Chromatography

Both thermospray and filament-on thermospray work best with solvents flowing at *ca.* 1 mL min^{-1} and so are ideal for coupling to standard-bore, on-line liquid chromatography. On-line separations offer advantages over discrete, loop injections for all samples other than the very pure ones. Chromatographic separation prior to mass spectrometry means (hopefully) that each chromatographic peak contains only one component and so all the ions in the spectrum relate to this one compound. This helps spectral interpretation, especially if the compound/mobile phase mixture generates many molecular-related ions (*e.g.* Table 6.3) which may easily be misassigned as originating from a number of different components rather than all arising from the same component. Chromatographic mass spectrometric separations also present the opportunity for solvent-generated background ions to be subtracted from the sample spectra, which helps to improve significantly the signal-to-noise ratio for the ions originating from the lower molecular mass components.

There are disadvantages of on-line chromatography too; some columns, mobile phase systems, and separations may not be compatible with the continual presence of ammonium acetate, and so an extra HPLC solvent delivery pump and T-piece may be necessary to add aqueous ammonium acetate to the mobile phase post-column [Figure 6.6(c)]. Many separations rely on gradient elution, and as the optimum thermospray operating conditions change substantially with respect to the mobile phase composition, the change in solvent can be a possible cause of instability and lack of sensitivity. These areas can be addressed, however, and some very elegant HPLC–MS experiments have been performed successfully.

Most 'real-life' separations of organic compounds involve gradient elutions that start with a high percentage of water in the mobile phase and end with a high percentage of an organic solvent such as methanol or acetonitrile. Therefore the optimum temperature of the capillary from which the mobile phase is spraying will drop significantly and steadily during the analysis, and during the optimization process, the optimum temperatures at these two extremes should be established. Now the choice of which temperature to use throughout the gradient elution has to be made. One solution is to choose a temperature that is in between the two optima, and hope that this compromise will not lead to too great a drop in sensitivity at the beginning and end of the run. With some mass spectrometry systems, a temperature ramp can be programmed to run simultaneously with the acquisition, or if this is not possible then at least the temperature can be changed manually during the run, although this latter solution is labour-intensive and, depending on the diligence of the operator, may not be reproducible from one analysis to the next. Another way to circumvent the optimization problem is to add aqueous ammonium acetate solution to the mobile phase after the column; for example if the gradient elution is proceeding at 1 mL min^{-1} (either with or without ammonium acetate present), then a make-up flow of 100 mM aqueous ammonium acetate can be introduced at the T-piece into this eluent at a rate of 0.5 mL min^{-1} [Figure 6.6(c)]. With this set-up, even though the gradient through the column is changing, there is always a significant amount of water and electrolyte entering the ionization source, and the optimum temperature will have less dependency on the gradient elution and will remain more or less constant throughout the entire run. Sometimes it has been necessary to add an organic solvent post-column to maintain spectral quality when the aqueous content of the mobile phase has been too high to generate acceptable spectra for the particular samples (in this case glutathione conjugates) in question.[15]

In summary, post-column addition is useful when the final solvent composition of the gradient contains too high an organic content to solubilize the electrolyte, when the aqueous content is too high for the samples in question, when the presence of an electrolyte is incompatible with the separation, or when the post-column addition of reference materials is necessary for accurate mass measurements.[14]

However, an increase in the organic content of the mobile phase has been shown to lead to a decrease in sensitivity in general,[16] and many such analyses have been performed with the filament **ON**, even if an electrolyte is present, to maintain the required level of sensitivity throughout the entire run. In an analysis such as this, presumably thermospray is the initial mode of ionization, and somewhere along the line, when the aqueous content of the mobile phase has decreased sufficiently, filament-on ionization takes over. Although the exact method of producing ions at any one point during the gradient may be under some debate, the important criteria of sensitivity and reproducibility are dealt with successfully, and good results can be obtained.

After a few years of blood, sweat and tears, from individuals and manufacturers alike, thermospray emerged with flying colours as the flagship of the pharmaceutical industry for both qualitative and quantitative drug analyses,[1,17] and also made a significant impact in other application areas such as environmental[2,4] and medical[3,6] fields. As a replacement for conventional GC–MS, this technique had the great advantage of being able to deal *directly* with polar compounds, thus avoiding time-consuming derivatization techniques.

Some of the most elegant thermospray ionization analyses in the pharmaceutical field have been in drug metabolism studies. These have tended to use reverse-phase mobile phases in conjunction with 4.6 mm or 3.9 mm i.d. HPLC columns at a solvent flow rate of 1–1.5 mL min^{-1}, with ammonium acetate (50–100 mM) often present throughout the separation as well as the filament being switched on, and an ultra-violet (UV) detector situated on-line to

[15] M. F. Bean, S. L. Pallante-Morell, D. M. Dulik and C. Fenselau, *Anal. Chem.*, 1990, **62**, 121.
[16] D. J. Liberato and A. L. Yergey, *Anal. Chem.*, 1986, **58**, 6.
[17] W. J. Blanchflower and D. G. Kennedy, *Biomed. Environ. Mass Spectrom.*, 1989, **18**, 935.

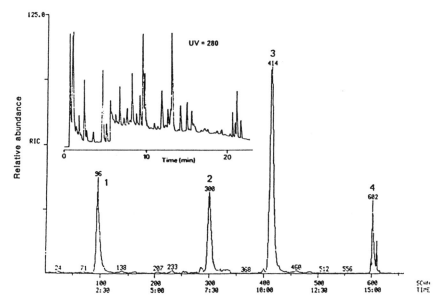

Figure 6.7 *UV trace (upper) and summed selected ion chromatogram thermospray trace (lower) of extracted rat faeces after dosage with the drug*
[Reproduced with permission from Wiley, ref. 1(b), fig. 7]

produce a chromatogram simultaneous with, and hence directly comparable with, the total ion chromatogram originating from the mass spectrometer.

In a typical study, the system is first tuned up on the drug of interest using the mobile phase required for the separation so that the optimum operating conditions are determined. Quite dirty matrices such as urine extracts can be injected; sometimes the first two or three minutes of the run, which can contain substantial amounts of endogenous material, is diverted away from the mass spectrometer to avoid contaminating the ionization source more than necessary. As thermospray produces predominantly molecular mass information, these analyses are frequently performed on multiple analyser mass spectrometers such as a tandem quadrupole or a magnetic sector–quadrupole hybrid instrument, so that further investigations can be undertaken making use of these MS–MS facilities to selectively generate structurally useful fragment ions in order to aid compound identification.

An example of such a run is shown in Figure 6.7. This run was acquired in positive ionization mode using a gradient elution of 15–45% aqueous acetonitrile with 100 mM ammonium acetate added flowing at 1.4 mL min^{-1} through a 3.9 mm i.d. reverse-phase HPLC column.[1b] The temperature of the capillary vaporizer was decreased during the run to accommodate the decrease in water content of the mobile phase. An extract from rat faeces after dosage with the drug was injected and eluted to generate the chromatograms shown. The UV trace is complex owing to endogenous materials, but four major components,

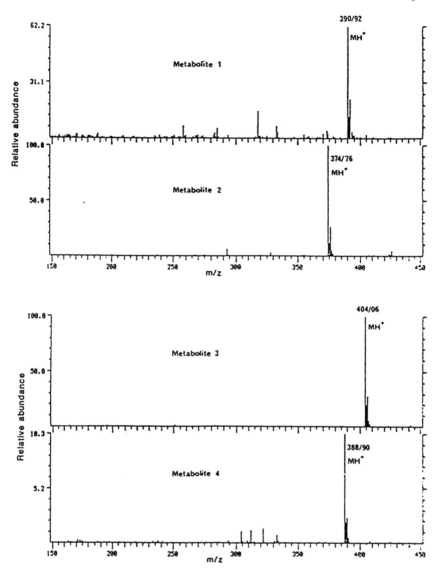

Figure 6.8 *Positive ionization thermospray spectra of the four metabolites from rat faeces*
[Reproduced with permission from Wiley, ref. 1(b), fig. 8]

whose spectra are depicted in Figure 6.8, are observed using mass spectral detection. From the four spectra, the molecular masses can be deduced as all four components have generated MH⁺ ions. It can be seen also from the isotope patterns of the MH⁺ ions that some ^{14}C labelling has been used for drug administration purposes. This provides extra confirmation that these ions are drug related. However, MS–MS is needed to generate structural informa-

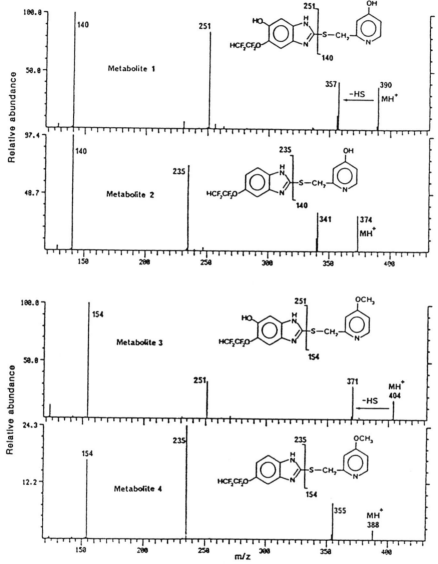

Figure 6.9 *Positive ionization thermospray MS–MS product ion spectra of the four metabolites from rat faeces*
[Reproduced with permission from Wiley, ref. 1(b), fig. 9]

tion. The product ion spectra shown in Figure 6.9 were all obtained by passing the MH⁺ ions individually through the first analyser of the mass spectrometer, generating fragments from these by collision with a suitable gas in the collision cell, and monitoring the resulting fragment ions using the second analyser of the mass spectrometer. The structures shown alongside these spectra have been labelled to indicate the fragmentation taking place. So in this study, both molecular mass and structural information have been obtained on several drug-related components as a result of a direct analysis from a dirty matrix without any prior compound isolation.

Quantitative analyses can also be performed with thermospray ionization, and detection limits as low as 2 ng mL^{-1} with acceptable coefficients of variation have been reported, for example in the case of clenbuterol.[17]

Finally, thermospray has been applied widely to pesticide analyses. Often in such cases, the presence of an ammonium acetate electrolyte will degrade the chromatography and so post-column addition or filament-on operation[2] is necessary. Detection limits have been reported in the region 1–18 μg L^{-1} for a number of pesticides.[4]

Supercritical Fluid Chromatography

Supercritical fluid chromatography (SFC) has been compared to both gas chromatography and high performance liquid chromatography because the physical properties of a supercritical liquid are similar in viscosity to a gas, but resemble a liquid in density.[18] SFC has been found to be useful in the analysis of non-volatile and thermally labile compounds, and both capillary and packed column SFC have been coupled successfully to mass spectrometry: the former, as the name suggests, makes use of capillary columns and so is well suited for GC–MS interfacing (see Section 4 of Chapter 3), while the latter has been used with a thermospray probe and ionization source, and will be discussed briefly here.

Packed column SFC compares favourably with HPLC in terms of providing faster analysis times and improved chromatographic resolution. For these reasons, it aroused a good deal of interest when thermospray was at its height of popularity, although only a few groups invested sufficient effort to create workable systems that solved real analytical problems.

Successful packed column SFC–MS analyses have been carried out using standard thermospray sources in filament-on mode, together with thermospray probes with the vaporizer tip crimped somewhat in order to maintain SFC conditions right up to the point where the eluent leaves the capillary.[18-20] Very briefly then, for packed column SFC–MS operation a functioning mass spectrometer equipped with a thermospray source with either filament-on ionization or a discharge electrode is required. An HPLC pump, a suitable injector, a conventional HPLC column (4.6 mm to 2 mm i.d.), and some means of heating the column are also needed, as is a supply of the gaseous mobile phase.

It is also advisable for the operator to have had some SFC experience before coupling SFC to a mass spectrometer.

[18] D. E. Games, A. J. Berry, I. C. Mylchreest, J. R. Perkins and S. Pleasance, *Eur. Chromatogr. News*, 1987, **1**, 10.

[19] J. R. Chapman, *Rapid Commun. Mass Spectrom.*, 1988, **2**, 6.

[20] W. M. A. Niessen, R. A. M. van der Hoeven, M. A. G. de Kraa, C. E. M. Heeremans, U. R. Tjaden and J. van der Greef, *J. Chromatogr.*, 1989, **474**, 113.

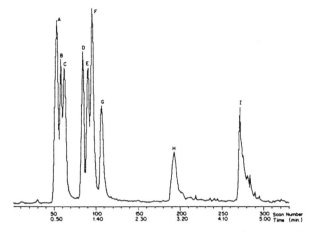

Figure 6.10 *SFC–MS analysis of a mixture of ten PTH-amino acids. The total ion chromatogram shows: A, Leu; B, Pro; C, Val; D, Ala; E, Met; F, Phe; G, Gly; H, Trp and Tyr; I, Asn.*
[Reproduced with permission from *Eur. Chromatogr. News*, ref. 18, fig. 4(a)]

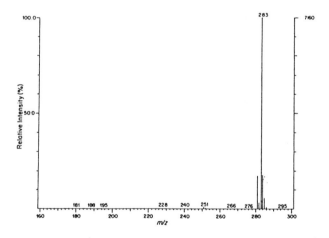

Figure 6.11 *SFC–MS analysis of a mixture of ten PTH-amino acids. Filament-on SFC–MS spectrum of peak F, PTH-Phe*
[Reproduced with permission from *Eur. Chromatogr. News*, ref. 18, fig. 5(a)]

Carbon dioxide (CO_2) is generally the supercritical mobile phase gas of choice due to the fact that it has a low critical temperature and pressure, and also it is non-flammable, non-toxic, and relatively inexpensive. The success of the chromatographic separation depends on the operating pressure of the mobile phase. In SFC, solute retention decreases with increasing density (or pressure, as density = pressure ÷ volume) as the solvating power of the supercritical liquid increases. So at a low supercritical liquid pressure, the solute retention times are longer, viscosities are lower, and separation efficiencies better than at higher supercritical liquid pressures. A range of mobile phase strengths can be achieved by the addition of a polar modifier, such as methanol, to the supercritical CO_2 mobile phase.

A typical mobile phase for SFC–MS analyses is CO_2 with 0–15% methanol added, at a flow rate of 1.5–2 mL min^{-1}.[20] For this mobile phase, a source temperature of 150 °C, a vaporizer temperature of 50 °C, and a discharge of 700 V were used for successful operation. With a system such as this, the reagent gas spectrum will consist mainly of ions at m/z 44 and 88, corresponding to the molecular ion of CO_2 and the $(CO_2)_2$ cluster ion respectively. If methanol is present as a modifier, then protonated methanol cluster ions will also be observed, *e.g.* at m/z 33, 65, 97 *etc.*, corresponding to $(CH_3OH)_nH^+$, where n = 1, 2, and 3 respectively. The presence of these ions, and of any others pertaining to the mobile phase in use, should be verified and optimized.

An example of packed column SFC–MS is presented in Figure 6.10.[18] The chromatogram shows the separation of ten PTH-amino acids using a standard reverse-phase HPLC column (4.6 mm × 10 cm) which was maintained at 75 °C. CO_2 was employed as the supercritical fluid mobile phase, to which the addition of methanol was programmed from 5% to 20% over 7 min. The mobile phase flow rate was maintained at 3 mL min^{-1} for one minute, and then raised to 4 mL min^{-1} for the remainder of the run. Although two of the PTH-amino acids are not resolved (PTH-Trp and PTH-Tyr), the peak shape is, in general, very good and the other components cleanly resolved. A spectrum from component F, PTH-Phe, is illustrated in Figure 6.11 and clearly shows predominant MH$^+$ ions at m/z 283. The other PTH-amino acids analysed in this run also gave spectra consisting almost exclusively of MH$^+$ ions.

The generation of molecular ions has also been noted and studies have been carried out introducing chemical ionization reagents to modify spectra, an example being the coaxial addition of diethylamine to the mobile phase in a sugar analysis to produce $[M + (C_2H_5)_2NH_2]^+$ ions.[19]

For packed column SFC–MS operation using a thermospray source, the mass spectrometer can be calibrated in standard thermospray mode before acquiring SFC–MS data.

CHAPTER 7

Matrix Assisted Laser Desorption Ionization

1 What Type of Compounds can be Analysed by Matrix Assisted Laser Desorption Ionization?

Matrix assisted laser desorption ionization (MALDI) is one of the more recent techniques developed with the aim of extending mass spectrometry to the analysis of large molecules with which the earlier techniques such as electron impact could not cope. In short MALDI deals excellently with thermolabile, non-volatile organic compounds, and it has been used very successfully for the analysis of biopolymers, such as proteins,[1] glycoproteins,[2] oligosaccharides,[3,4] and oligonucleotides,[5] where its ease of use and its tolerance to the presence of buffers has guaranteed its widespread acceptance. In addition MALDI is probably the most effective method of analysing synthetic polymers,[6,7,8] producing not only molecular mass information but also end-group characterization.

2 Matrix Assisted Laser Desorption Ionization

The Principles of Matrix Assisted Laser Desorption Ionization

In order to determine the molecular weight of an intact polymer or biopolymer by mass spectrometry, the sample needs to be converted into intact molecular ions, quite a difficult task when dealing with large, polar, involatile samples. Several earlier approaches to the ionization of such compounds have emerged over the years, including field desorption (see Chapter 5) which makes use of

[1] F. Hillenkamp, M. Karas, R. C. Beavis and B. T. Chait, *Anal. Chem.*, 1991, **63**, 1193.
[2] C. W. Sutton, J. A. O'Neill and J. S. Cottrell, *Anal. Biochem.*, 1994, **34**, 218.
[3] D. J. Harvey, P. M. Rudd, R. H. Bateman, R. S. Bordoli, K. Howes, J. B. Hoyes and R. G. Vickers, *Org. Mass Spectrom.*, 1994, **29**, 753.
[4] D. Garrozzo, G. Impallomeni, E. Spina, L. Sturiale and F. Zanetti, *Rapid Commun. Mass Spectrom.*, 1995, **9**, 937.
[5] A. Overberg, A. Hassenburger and F. Hillenkamp, in *Mass Spectrometry in Biological Sciences: A Tutorial*, ed. M. L. Gross, Kluwer Academic, Amsterdam, 1992, p. 181.
[6] U. Bahr, A. Deppe, M. Karas and F. Hillenkamp, *Anal. Chem.*, 1992, **64**, 2866.
[7] R. S. Lehrle and D. S. Sarson, *Rapid Commun. Mass Spectrom.*, 1995, **9**, 91.
[8] G. Montaudo, M. S. Montaudo, C. Puglisi and F. Samperi, *Rapid Commun. Mass Spectrom.*, 1995, **9**, 453.

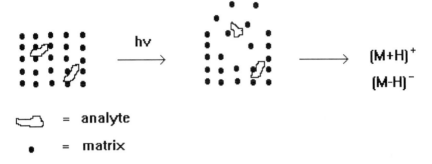

Figure 7.1 *Matrix assisted laser desorption ionization (MALDI)*

an electric field, fast atom or ion bombardment (see Chapter 4) where a beam of high energy atoms or ions is directed at the sample, electrospray ionization (see Chapter 2) in which sample ions evaporate from small, highly charged droplets, and recently laser desorption ionization (LDI),[1] the subject of this Chapter, which involves sample bombardment with short, intense pulses from a laser light to effect both the desorption and ionization of the molecules.

Many different types of laser light have been used, covering a range of wavelengths and pulse widths. The ideal laser should deliver an efficient and controllable quantity of energy to the sample, and in order to avoid thermal decomposition this energy must be transferred quickly. Early studies employing laser focusing directly onto the sample indicated that very high molecular mass samples (> 9000 da) could not be ionized in their intact form and sometimes results were not reproducible, but later work demonstrated that these problems could be overcome by mixing the substrate with a highly absorbing matrix,[9] thus separating out the desorption and ionization steps. The technique became known as matrix assisted laser desorption ionization. From the array of available ionization techniques, the two currently favoured ones for high mass analyses are matrix assisted laser desorption ionization and electrospray, which together have largely superseded the other, earlier methods.

The key to success in MALDI operation seems to be in having a low concentration of sample molecules in the matrix (Figure 7.1). The matrix transforms the laser energy into excitation energy for the sample which leads to sputtering of the analyte and matrix from the surface of the mixture. In this way the energy transfer is efficient, while the sample molecules are spared from excessive energy which would under other circumstances lead to their decomposition. It is thought also that suitable matrices enhance the substrate ion formation by a process of photoexcitation or photoionization of the matrix followed by proton transfer to or from (for positive or negative ionization

[9] M. Karas, D. Bachmann, U. Bahr and F. Hillenkamp, *Int. J. Mass Spectrom. Ion Processes*, 1987, **78**, 53.

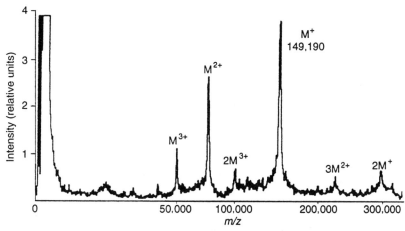

Figure 7.2 *MALDI spectrum of a monoclonal mouse antibody of molecular weight 149 190 da, using nicotinic acid matrix and a laser wavelength of 266 nm* (Reprinted with permission from *Anal. Chem.*, ref. 1)

respectively) the substrate molecule.[1] An example of a MALDI analysis for a high molecular mass sample is shown in Figure 7.2, where a sample of a monoclonal mouse antibody was analysed using nicotinic acid as matrix. MH^+ ions of the intact protein are clearly apparent at *m/z* 149 190, together with some multiply charged and polymeric ions also arising from the intact molecule but of weaker intensity. The wavelength of the laser in this example was 266 nm, and a time-of-flight (TOF) mass spectrometer was employed to analyse the ions generated.

A TOF mass analyser is usually interfaced to laser desorption ionization techniques due to the virtually unlimited mass range available and its ability to cope with the short time scale of pulsed laser ionization. A TOF instrument can record simultaneously all the ions produced from each single laser pulse, whereas a scanning mass spectrometer such as one with a magnetic sector or a quadrupole analyser requires an uninterrupted presence of sample ions to be able to scan continuously through all of the appropriate *m/z* range. A simplified schematic diagram of a linear (*i.e.* the ions have a straight flight path) MALDI–TOF instrument is shown in Figure 7.3. The packet of ions produced by the laser pulse is extracted and accelerated by an electric field (up to 30 kV), and then drifts through the field-free region to the detector. The flight path length is typically between 0.1 m and 3 m, and the velocity of a particular ion is proportional to its mass-to-charge ratio. The time of arrival of ions with a common start time at the detector is thus of importance as it is used to calculate the masses of the various ions.

Most types of commercial MALDI–TOF mass spectrometers now have a pulsed N_2 laser of wavelength 337 nm, and this is the type generally referred to in the following Sections of this Chapter.

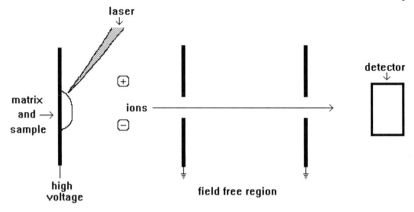

Figure 7.3   *Matrix assisted laser desorption ionization (MALDI) time-of-flight (TOF)*
mass spectrometry (linear mode)

Until recently, the ability of the linear TOF mass spectrometer to produce
separate signals from ions of similar mass was generally acknowledged to be
less than a magnetic sector or a quadrupole mass spectrometer. One reason for
this was that ions of the same mass-to-charge ratio had widely differing
energies, which resulted in the broad peak widths observed at the detector, and
in order to compensate for this most TOF mass spectrometers are manufactured with a reflectron placed in the flight path (Figure 7.4; see also Chapter 1,
Section 2). In this situation, the higher energy ions penetrate further into the
reflectron than do the lower energy ones, and when they all emerge from the
reflectron (often after a 180° change in direction), the energy differences have
been corrected to some extent with the result that the resolution and mass
accuracy of the mass spectrometer is increased. Many commercial instruments
have the facility for both linear and reflectron analyses.

A comparison between spectra acquired using linear and reflectron TOF
mass spectrometry is shown in Figure 7.5. The molecular ion region is
highlighted for the peptide ACTH (clip 18–39) and shows reasonable resolution of the isotopes in the case of the reflectron TOF analysis, compared with a
single, averaged peak at *m/z* 2466.7 in the case of the linear TOF analysis. The
mass accuracy using the reflectron mode can be expected to be as good as
0.01%, which is a factor of ten better than that expected when operating in the
linear mode.

A further technique now applied to MALDI–TOF mass spectrometry
employs 'delayed extraction'[10,11] of the ions exiting the ionization source, and
this also helps minimize small differences in ion energies and thus improves
significantly resolution, signal-to-noise and mass accuracy. For example, the

[10] W. C. Wiley and I. McLaren, *Rev. Sci. Inst.*, 1955, **26**, 1150.
[11] J. J. Lennon and R. S. Brown, *Annal. Chem.*, 1995, **67**, 1988.

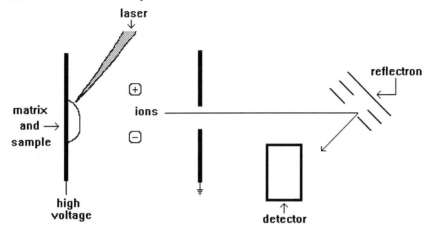

Figure 7.4 *Matrix assisted laser desorption ionization (MALDI) time-of-flight (TOF) mass spectrometry (reflectron mode)*

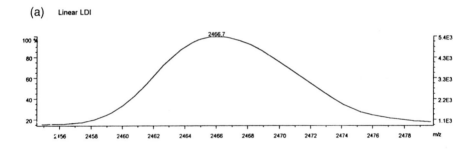

Figure 7.5 *MALDI spectra of the peptide ACTH (clip 18-39) acquired in (a) linear TOF mode; and (b) reflectron TOF mode*
(Reprinted with permission from Micromass UK Ltd.)

Table 7.1   *Popular MALDI matrices for use at λ = 337 nm, their application areas, and some commonly used solvents*

MALDI matrix	Average MW da	Applications	Solvents
3-amino-4-hydroxybenzoic acid	153.1	saccharides, oligosaccharides	acetonitrile, water
cinnamic acid	148.1	general	water, acetonitrile
α-cyano-4-hydroxycinnamic acid	189.2	peptides, lipids, nucleotides, oligonucleotides, polymers	water, acetonitrile, ethanol, acetone
2,5-dihydroxybenzoic acid (DHB)	154.1	saccharides, oligosaccharides, peptides, nucleotides, oligonucleotides, polymers	water, acetonitrile, chloroform, acetone, methanol
6,7-dihydroxycoumarin (esculetin)	178.2	peptides, lipids	water
3,5-dimethoxy-4-hydroxycinnamic acid (sinapinic acid)	224.2	proteins, peptides, polymers	water, acetonitrile, chloroform, acetone
dithranol	226.2	polymers	tetrahydrofuran
2-(4-hydroxyphenylazo)benzoic acid (HABA)	242.2	polymers	acetone
3-hydroxypicolinic acid	139.1	oligonucleotides	ethanol
3-β-indole acrylic acid	187.2	polymers	acetone
5-hydroxy-2-methoxybenzoic acid	168.2	lipids	water
nicotinic acid (absorbs at 266 nm)	123.1	proteins	water
2,4,6-trihydroxyacetophenone	168.2	oligonucleotides	ethanol

resolution of a modern TOF instrument could be as high as 12 000 (FWHM). Delayed extraction techniques (see also Chapter 1, Section 2) are particularly useful for labile molecules such as oligonucleotides because sample–matrix collisions are minimized which leads to reduced chemical noise and enhanced spectra.

Practical Operation of Matrix Assisted Laser Desorption Ionization

Essential Requirements for Operation

The following list includes some of the essential requirements for MALDI–TOF mass spectrometry:

Table 7.2 *Application areas and their appropriate matrices for MALDI analyses at λ = 237 nm*

Application	Matrix
peptides	α-cyano-4-hydroxycinnamic acid
	2,5-dihydroxybenzoic acid
	6,7-dihydroxycoumarin
	3,5-dimethoxy-4-hydroxycinnamic acid
proteins	3,5-dimethoxy-4-hydroxycinnamic acid
oligosaccharides	3-amino-4-hydroxybenzoic acid
	2,5-dihydroxybenzoic acid
oligonucleotides	α-cyano-4-hydroxycinnamic acid
	2,5-dihydroxybenzoic acid
	3-hydroxypicolinic acid
	2,4,6-trihydroxyacetophenone
lipids	α-cyano-4-hydroxycinnamic acid
	6,7-dihydroxycoumarin
	5-hydroxy-2-methoxybenzoic acid
polymers	5-chloro-2-hydroxybenzoic acid
	5-dihydroxybenzoic acid
	3,5-dimethoxy-4-hydroxycinnamic acid
	dithranol
	2-(4-hydroxyphenylazo)benzoic acid

- a functioning time-of-flight mass spectrometer under vacuum as detailed in the manufacturer's manuals, equipped with a MALDI source and laser;
- MALDI targets on which to load the sample;
- a selection of matrices (see Chapter 7, Tables 7.1 and 7.2);
- a range of suitable calibrants (see Chapter 7, Table 7.3);
- a range of sample vials, pipettes, and syringes for sample preparation;
- a range of solvents and additives for sample preparation.

Choice of Matrix

The matrix is present to absorb light from the laser and isolate the analyte molecules from each other, therefore the basic requirements of a matrix material are that it should absorb at the wavelength of the laser, and that it should be present in an excess of *ca.* 5000:1 to the sample to be analysed.

A range of MALDI matrices which are suitable for use with a laser light of wavelength (λ) 337 nm is listed in Table 7.1, together with the general application areas and some of the solvents that are commonly used to dissolve them. It can be seen that there is usually more than one matrix recommended for each application area, and the operator should make an initial investigation by trying several matrices with a sample similar to the ones requiring analysis. This way the user can make a personal choice for the particular compounds under scrutiny. The same matrices have also been summarized according to their application areas in Table 7.2.

It should also be remembered that the lists supplied here are by no means exhaustive ones, they are simply the most *commonly* encountered ones in the literature, and there is no reason at all why 'new' matrices should not be tried out. More details are supplied for the use of specific matrices in various application areas later on in this Section.

Generally, the matrix solutions should be prepared each day as they are susceptible to light-induced decomposition. A typical solution would comprise the matrix at a concentration of 10 mg mL^{-1} in a solvent system that is compatible with the samples to be analysed, and can include *inter alia* water, ethanol, acetonitrile, acetone, chloroform, tetrahydrofuran, and mixtures thereof. Often at such a high concentration, the matrix solution is saturated and needs to be allowed to settle after thorough mixing has taken place.

The matrix 3,5-dimethoxy-4-hydroxycinnamic acid, for example, can be used for the analysis of both proteins and synthetic polymers, two compound classes of widely differing polarities. For the protein analyses the matrix works well when dissolved in a solvent mixture of 7:3 (*v/v*) water–acetonitrile with 0.1% trifluoroacetic acid added, while for the analysis of less polar, synthetic polymers then non-aqueous organic solvents such as chloroform or acetone are more appropriate for dissolution of the matrix. Additives are often mixed with the matrix solution to promote ionization and produce higher ion yields, and several specific examples are illustrated for different compound classes in the following text of this Section.

Sample Preparation and Analysis[12]

The sample to be analysed should be dissolved in an appropriate solvent at a concentration of 10–50 pmol μL^{-1} and an aliquot (1–2 μL) of this removed and mixed in a small sample vial with an equal volume of the matrix solution. As the matrix solution is at a concentration of *ca.* 10 mg mL^{-1}, this results in an excess of matrix compared to the sample. An amount (1–2 μL) of the final solution is removed and applied to the sample target, which should be washed well with solvents and then allowed to dry prior to use. The operator should avoid touching the sample target as this will deposit contaminants on the surface. New targets should be used whenever possible, as thorough washing of the target is not always successful in removing old deposits which may then interfere with the spectrum of the new sample. The sample/matrix solution is allowed to dry on the target at room temperature and pressure for several minutes, and then the target is inserted into the source of the mass spectrometer. With the sample target in position and a good vacuum achieved, data can be accumulated until an adequately intense spectrum has been amassed, from which an approximate (uncalibrated) molecular weight can be obtained. A reasonably intense solution will require typically 30–50 laser shots per spectrum.

---

[12] J. A. Carroll and R. C. Beavis, in 'Matrix-Assisted Laser Desorption and Ionization', *Laser Desorption and Ablation*, eds., J. C. Miller, R. F. Hagulund, ch. 7, 1996 (Series: Experimental Methods in the Physical Sciences).

Table 7.3 *Useful MALDI calibrants*[13]

Calibrant	Average MW da	Solution
gramicidin-S	1141.5 (1140.7059 monoisotopic)	0.1% aqueous trifluoroacetic acid
ACTH (18–39 clip)	2465.7 (2464.1910 monoisotopic)	0.1% aqueous trifluoroacetic acid
insulin (bovine)	5733.5 (5729.6009 monoisotopic)	0.1% aqueous trifluoroacetic acid
cytochrome-C (horse heart)	12 360	0.1% aqueous trifluoroacetic acid
trypsinogen	23 980	0.1% aqueous trifluoroacetic acid
serum albumin (bovine)	66 430	0.1% aqueous trifluoroacetic acid

A standard commercial MALDI–TOF mass spectrometer will probably have a nitrogen laser which produces a 4 ns pulse at a wavelength of 337 nm with each pulse having a fixed energy of 180 μJ and the energy arriving at the sample/matrix surface can be optimized for each sample. The matrix absorbs the energy of the laser and then ionizes the sample, and the resulting ions are accelerated out of the source, which is held at *ca.* $\pm$ 25 kV, depending on whether positive or negative ions are to be detected.

When the sample has been analysed, it is necessary to calibrate the mass scale. An external calibration (*i.e.* the calibrant is analysed on a separate sample target) or an internal calibration (*i.e.* the sample is re-analysed with the calibrant mixed in with the sample on the same sample target) can be performed, the latter producing an improved mass accuracy of a factor of *ca.* 10, compared with the former method. A number of samples suitable for calibration purposes and which will cover the molecular weight range of the samples under investigation should be prepared and stored, in the refrigerator if necessary, so that they are easily accessible for use. In each case a concentration of *ca.* 10 pmol μL^{-1} in water with 0.1% trifluoroacetic acid present is suggested, and Table 7.3 presents examples of samples suitable for calibration.[13]

Once the sample has been analysed, the most appropriate calibrant can be chosen. If an external calibration is required then the calibrant (or mixture of calibrants) is analysed and a two point calibration generated, from which the sample spectrum can be recalibrated. The two points used in a MALDI–TOF calibration are usually the protonated (or deprotonated) molecular ion of a calibrant together with either its dimer or doubly charged equivalent, or else the protonated (or deprotonated) molecular ions from calibrants which have been mixed together. If an internal calibration is required for better mass

[13] W. G. Critchley, Micromass UK Ltd., Manchester, UK, personal communication.

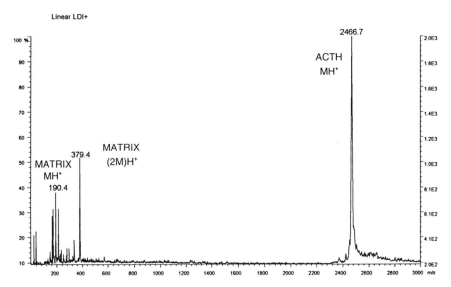

Figure 7.6 *MALDI spectrum of the peptide ACTH (clip 18-39) acquired in linear TOF mode using α-cyano-4-hydroxycinnamic matrix, showing both peptide and matrix ions*
(Reprinted with permission from Micromass UK Ltd.)

accuracy, the sample solution should be mixed with an equimolar amount of the calibrant and the usual quantity of matrix solution, then respotted onto another sample target and reanalysed. In these cases it is important that the calibrant does not generate any ions which could interfere with the spectrum generated by the sample. A two point calibration can then be generated and the same spectrum recalibrated. The matrix will also produce MH⁺ ions (see Table 7.1 for the molecular weights of the matrices), and if these are not saturated, they too can be employed for calibration purposes.

TOF data corresponding to low energy MS–MS product ion experiments can be acquired on some instruments by monitoring the Post Source Decay (PSD)[14,15] of selected precursor ions. The low energy dissociation of these precursor ions in the TOF mass analyser gives rise to metastable ions that can be detected by changing the reflectron voltage according to the *m/z* values of the fragment ions.

The above information is applicable to all types of sample analyses. Special mention must be made, however, for some of the more popular MALDI–TOF application areas, *viz.* peptides and proteins, oligosaccharides, and synthetic polymers.

[14] R. Kaufmann, B. Spengler and F. Lützenkirchen, *Rapid Commun. Mass Spectrom.*, 1993, 7, 902.
[15] B. Spengler in 'New Instrument Approaches to Collision Induced Dissociation using a TOF Instrument', *Protein and Peptide Analysis by Mass Spectrometry*, ed. J. R. Chapman, Humana Press Inc., NJ, 1996.

Table 7.4 *Concentrations of common contaminants tolerated by MALDI–TOF protein analyses*[13]

Contaminant	Maximum concentration tolerated
phosphate buffer	20 mM
TRIS buffer	50 mM
detergents	0.1%
sodium dodecyl sulfate (SDS)	0.01%
alkali metal salts	1 M
glycerol	2%
ammonium hydrogen carbonate	30 mM
guanidine	1 M
sodium azide	1%

Peptides and Proteins. Most of these samples are water soluble and can be dissolved in 0.1% aqueous trifluoroacetic acid and analysed in the positive ionization mode. The presence of the acid aids protonation of the peptide or protein sample, and in general the spectrum is dominated by intense MH^+ ions, accompanied by weaker, doubly charged $[(M + 2H)^{2+}]$ and dimeric $[(2M + H)^+]$ species.

Samples up to 12 000 da molecular weight tend to produce good results with the matrix α-cyano-4-hydroxycinnamic acid. If using this matrix, most samples will produce intense MH^+ ions when the matrix solution is 1:1 (*v/v*) water–acetonitrile with 0.1% trifluoroacetic acid added, while increasing the organic content of the matrix solution will promote the formation of multiply charged ions at the expense of the singly charged, protonated molecular ion. Figure 7.6 shows the spectrum of the peptide ACTH (18–39 clip), which is often used as a calibrant, using α-cyano-4-hydroxycinnamic acid as matrix. The spectrum shows not only the MH^+ ions corresponding to the peptide at *m/z* 2466.7, but also MH^+ and $(2M + H)^+$ ions corresponding to the matrix at *m/z* 190.4 and 379.4.

Samples above 12 000 da molecular weight usually generate good results with 3,5-dimethoxy-4-hydroxycinnamic acid (sinapinic acid) matrix which has been dissolved in 6:4 (*v/v*) water–acetonitrile with 0.1% trifluoroacetic acid added. In some cases, sinapinic acid adducting is observed in addition to the MH^+ ions. These adducts are apparent at $(M + 206)^+$ and correspond to the addition of sinapinic acid accompanied with dehydration.

Protein samples are often contaminated with both inorganic and organic salts and buffers. MALDI–TOF copes well with these contaminants, significantly better than electrospray ionization does, and so these samples can be analysed without prior purification unless the degree of contamination is excessive. A list of common contaminants is displayed in Table 7.4, together with the maximum level that can be tolerated by MALDI–TOF and other, more detailed lists have been compiled.[16]

[16] O. Vorm, B. T. Chait and P. Roepstorff in 'Mass Spectrometry of Protein Samples containing Detergents', *Proc. 41st ASMS Conf.*, 621 (1994).

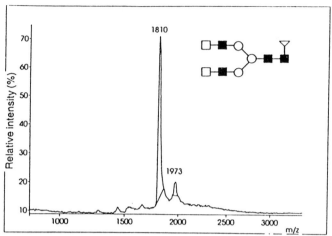

Figure 7.7 *MALDI spectrum of an N-linked sugar acquired in linear TOF mode using 2,5-dihydroxybenzoic acid matrix showing MNa⁺ ions*
(Reprinted with permission from *Org. Mass Spectrom.*, ref. 3)

Oligosaccharides. With saccharide samples, the matrix 3-amino-4-hydroxybenzoic acid or more preferably 2,5-dihydroxybenzoic acid[3,4] can be used in 8:2 to 3:7 (*v/v*) water–acetonitrile solutions. After mixing the sample solution (usually saccharides dissolve in water alone) with the 2,5-dihydroxybenzoic acid matrix solution, and then applying an aliquot of this combination to the sample target, a small amount of ethanol (0.5 μL) added to the dried target to recrystallize the sample and matrix has been found to produce a more uniform layer of crystals, with excellent incorporation of the sample into the matrix lattice, and hence greater sensitivity.[3]

In addition to the calibrants listed in Table 7.3, any oligosaccharide standard of which the molecular weight is known accurately and confidently can be used for calibration purposes.

In general, a higher energy is required for the analysis of saccharides, and the sensitivity of MALDI–TOF to saccharides has been reported to be good, more than ten times the sensitivity achievable with fast atom bombardment, for example.[17] In contrast to protein and peptide samples, oligosaccharide samples usually generate solely a sodium adduct (MNa⁺) of the molecular ion in positive ionization mode, with little evidence for multiply charged ions or dimeric species. An example of this is shown in Figure 7.7, the spectrum of an underivatized *N*-linked oligosaccharide of molecular mass 1787.6 da[3] accompanied by a small impurity of higher molecular mass. As can be seen, when using a linear TOF mass spectrometer in this way, there is little evidence of fragmentation, and as complex mixtures generally produce solely MNa⁺ ions for each component, the spectra are quite unambiguous to interpret.

[17] M. C. Huberty, J. E. Vath and S. A. Martin, *Anal. Chem.*, 1993, **65**, 2791.

Oligosaccharides (and glycopeptides) containing carboxylic acid functional groups produce good results using negative ionization MALDI–TOF, where $(M - H)^-$ ions are generally the dominant species.[3]

Oligonucleotides. Oligonucleotides typically produce excellent results by negative ionization MALDI–TOF analysis, and a suitable matrix to use is 2,4,6-trihydroxyacetophenone as a 0.5 M ethanolic solution. Oligonucleotides are generally sufficiently polar to be used as aqueous solutions. The addition of ammonium salts to promote ammoniated molecular ions which dissociate in the TOF analyser producing $(M - H)^-$ sample ions is recommended.[13] The ammonium ions, either as a 0.1 M aqueous solution of ammonium tartrate or a 0.1 M aqueous solution of ammonium citrate dibasic, are mixed with the matrix solution prior to the addition of the sample solution.

Synthetic Polymers. The field of synthetic polymers is another area in which MALDI is very successful. Until the advent of MALDI, field desorption was the method of choice for these polymers, at least those with moderate molecular weights which did not surpass the *m/z* ranges of the magnetic sector mass spectrometers. MALDI has been shown to be capable of analysing synthetic polymers such as polyesters, polyethers, polystyrenes, and acrylics, providing not only molecular weight information but also end group characterization. In general, results agree quite well with those obtained by gel permeation methods, although there is some debate about this accuracy at the moment.[7,8]

The procedure for analysing polymeric samples, including the choice of solvent and matrix, depends on the class of polymer under investigation. For example, polyethylene glycols have been dissolved in water, polypropylene glycols in a mixture of 1:1 (*v/v*) water–ethanol, and polymethylmethacrylates in acetone, each at a concentration of 5 g L^{-1}.[6] Another report favours dissolving polymethylmethacrylates in tetrahydrofuran at a concentration of 1 g L^{-1}.[7] Polystyrene and polyether samples have been dissolved in chloroform, and polyesters in tetrahydrofuran.[18]

A variety of matrices has been employed (see Table 7.1 for a list of matrices and appropriate solvents), the most popular including 2,5-dihydroxybenzoic acid [for polyethylene glycols, polypropylene glycols[6] and, with the matrix dissolved in 0.1% trifluoroacetic acid in 1:1 (*v/v*) water–ethanol, polymethylmethacrylates[7]], dithranol (for polystyrenes and polyesters[18]), 3-β-indole acrylic acid (for polymethylmethacrylates[18]) and 2-(4-hydroxyphenylazo)benzoic acid (for polyethylene glycols, polymethylmethacrylates, and polystyrenes[8]). If possible, it is recommended that the sample and matrix are dissolved in the same solvent system.

In many cases, salts have been added to the target after mixing the matrix

[18] J. H. Scrivens, H. T. Yates, M. J. Taylor, J. A. Segal, A. Jackson, A. J. Russell, A. K. Chaudhary, A. Diaf, E. J. Beckman, G. Critchley, A. E. Ashcroft, S. Campbell and T. Williams, *The Analysis of Synthetic Polymer Formulations by MALDI–TOF Mass Spectrometry,* Application Note 42, Micromass UK Ltd., Floats Road, Manchester, UK, 1995.

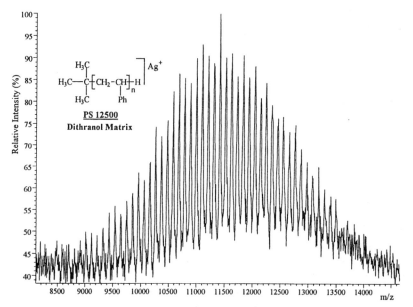

Figure 7.8 *MALDI spectrum of polystyrene 12 500 acquired in linear TOF mode using dithranol matrix showing MAg⁺ ions with a repeat polystyrene unit of 104 da* (Reprinted with permission from Micromass UK Ltd., ref. 18, fig. 5)

and sample solutions, in order to promote cationization. Often some degree of cationization will take place in the absence of additives, but almost exclusive adducting, and hence spectral simplification, can result from purposely adding small amounts of suitable salts. The recommended procedure is mixing an equal volume of a 0.1 M solution of a salt such as lithium iodide, sodium iodide, or potassium iodide in either water or acetone, or aqueous silver trifluoroacetate solution (5 g L^{-1}),[6] or silver trifluoroacetate dissolved in tetrahydrofuran at a concentration of 1 mg mL^{-1},[18] or a saturated solution of silver nitrate in chloroform[8] to the mixed matrix and sample solution before applying the final sample mixture to the target. Hence commonly observed adducts are MLi⁺ (M + 7 da), MNa⁺ (M + 23 da), MK⁺ (M + 39 da), and MAg⁺ (M + 108 da).

The MALDI–TOF spectrum of polystyrene 12 500 with dithranol as the matrix is shown in Figure 7.8. The repeat polystyrene unit of 104 da is apparent, and all individual oligomers are detected as the MAg⁺ ions due to the presence of silver trifluoroacetate mixed with the sample and the matrix.

A Comparison of MALDI–TOF and Electrospray

The two currently favoured ionization techniques for the analysis and identification of biomolecules are MALDI and electrospray. There are advantages with both techniques, although it is difficult to compare the two directly as they are usually affiliated with different types of mass spectrometer, and hence some of their characteristics result from the limitations or benefits of the type of mass analyser used.

As already stated, MALDI ionization is generally associated with a time-of-flight mass spectrometer, and electrospray with either a quadrupole or a magnetic sector mass spectrometer although recent developments have coupled electrospray to time-of-flight mass spectrometers. All of these analysers measure ions on a mass-to-charge ratio, but whereas MALDI generates intact quasi-molecular ions which require the large *m/z* range provided by a TOF analyser, electrospray produces multiply charged ions which can be measured on a typical quadrupole or magnetic sector mass spectrometer in the *m/z* region 500 to 3000. Thus the appearance of the spectra produced are quite different.

There are differences also in the method of sampling employed for the two ionization techniques; electrospray is an 'on-line' technique using a continuous flow of solvent and so is ideal for direct coupling with liquid chromatography and capillary electrophoresis, whereas MALDI samples from a solid target and this, coupled with its greater tolerance to buffers, makes it applicable to the analysis of samples from membranes, for example.

The molecular mass ranges for both MALDI and electrospray are high. A TOF mass spectrometer has an unlimited theoretical mass range, and in practice molecular weights have been recorded up to one million da. Electrospray does less well than MALDI with such *extremely* high molecular weight samples, its practical limitation being *ca.* 150 000 da, which is due to the complexity of the spectrum caused by the great number of multiply charged ions occurring in a relatively small region of the range *m/z* 500 to 3000.

Mass accuracy is always of importance, whether verifying the presence of a known compound or characterizing a new one. Electrospray does well in this respect, often exceeding a mass accuracy of 0.01% on samples up to *ca.* 40 000 da. This is usually good enough to verify the amino acid sequence of a known protein, and if the measured mass does not agree with the calculated one, the sequence may be deemed incorrect. MALDI, with the TOF analyser operating in linear mode, will give an accuracy of 0.5%, which can be improved with the use of an internal rather than an external calibrant. If a reflectron lens is present on the TOF analyser, then this accuracy can be further increased to 0.02%, or better. With delayed extraction facilities the mass accuracy, as well as the resolution, of a TOF analyser is further enhanced, and mass measurements of < 10 ppm are possible.

Both electrospray and MALDI can be used successfully to generate structural information for peptides, glycopeptides, saccharides, and oligonucleotides, if used in conjunction with some method of collision, induced MS–MS for electrospray or post-source decay for MALDI–TOF.

It should be remembered that both techniques are sensitive: femtomole (or even less) sensitivity is possible, both are suitable for mixture analysis, both have reasonably short analysis times and can be automated, and neither has been accredited with the difficulty of operation or 'black art' associated with earlier techniques such as FD/FI and FA/IB. The future for biochemists has never looked better!

Appendix 1

Some Common Abbreviations

amu	atomic mass unit (see da)
APCI	atmospheric pressure chemical ionization
API	atmospheric pressure ionization
B	magnetic analyser; field strength of a magnet
CE	capillary electrophoresis
CI	chemical ionization
da	dalton or atomic mass unit (see amu); $^{12}C = 12$ da
DCI	desorption chemical ionization
EI	electron impact
ES	electrospray
ESA	E; electrostatic analyser
eV	electron volt; 23.06 kcal mol^{-1}
FAB	fast atom bombardment
FD	field desorption
FI	field ionization
FIB	fast ion bombardment
GC	gas chromatography
IE	ionization energy
LC	liquid chromatography
LSIMS	liquid secondary ionization mass spectrometry
$M^{+\bullet}$	molecular ion
MALDI	matrix assisted laser desorption ionization
MH^+	protonated molecular ion
$(M-H)^-$	deprotonated molecular ion
mmu	millimass units; $10^{-3} \times$ amu
MS	mass spectrometry
MS–MS	mass spectrometry–mass spectrometry; tandem mass spectrometry
m/z	mass of an ion in daltons divided by its charge; Thompson
PA	proton affinity
PB	particle beam
PFK	perfluorokerosene
ppm	parts per million
PSP	plasmaspray
Q	quadrupole analyser
SFC	supercritical fluid chromatography

SIR	selected ion recording
TOF	time-of-flight mass spectrometer
TSP	thermospray
V	volt; accelerating voltage

Appendix 2

Some Common Reference Compounds

1 Caesium iodide–positive ions

Caesium iodide can be analysed in aqueous solution at a concentration of *ca.* 2 mg mL^{-1} by fast atom/ion bombardment and electrospray. The following singly charged, monoisotopic, $[Cs(CsI)_n{}^+]$ cluster ions are observed in positive ionization mode:

132.9054
265.8109
392.7153
652.5253
912.3352
1172.1451
1431.9550
1691.7649
1951.5748
2211.3847
2471.1946
2731.0045
2990.8144
3250.6244
3510.4343
3770.2442

etc., incrementing by 259.8099 da each time.

2 Caesium iodide–negative ions

Caesium iodide can be analysed in aqueous solution at a concentration of *ca.* 2 mg mL^{-1} by fast atom/ion bombardment and electrospray. The following singly charged, monoisotopic, $[(I(CsI)_n{}^-]$ cluster ions are observed in negative ionization mode:

126.9045
386.7144
906.3342
1166.1441
1425.9540
1685.7639
1945.5738
2205.3838
2465.1937
2725.0036
2984.8135
3244.6234
3504.4333
3764.2432
4024.0531

etc., incrementing by 259.8099 da each time.

3 Heptacosafluorotributylamine [heptacosa, FC43 $(C_4F_9)_3N$]

Heptacosa can be analysed in its neat form by electron impact and negative chemical ionization. The following singly charged, monoisotopic, ions are observed:

49.9938	225.9903
68.9952	230.9856
92.9952	263.9871
99.9936	313.9839
113.9967	325.9839
118.9920	375.9807
130.9920	413.9775
149.9904	425.9775
163.9935	463.9743
168.9888	501.9711
175.9935	537.9711
180.9888	575.9680
213.9903	613.9647
218.9856	

4 Horse heart myoglobin–positive ions

Horse heart myoglobin (MW 16 951.499) can be analysed by electrospray ionization at a concentration of 20 pmol μL^{-1} in 1:1 (v/v) acetonitrile–0.1% aqueous formic acid. The following multiply charged, averaged ions [(M + nH)$^{n+}$] are observed in positive ionization mode:

606.4186	998.1549
628.8412	1060.4766
652.9886	1131.1078
679.0679	1211.8293
707.3204	1304.9694
738.0296	1413.6328
771.5306	1542.0533
808.2221	1696.1578
848.5829	1884.5078
893.1921	2119.9453
942.7577	2422.6506

5 Horse heart myoglobin–negative ions

Horse heart myoglobin (MW 16 951.499) can be analysed by electrospray ionization at a concentration of 20 pmol μL^{-1} in 1:1 (v/v) water–propan-2-ol. The following multiply charged, averaged, ions [(M − nH)$^{n-}$] are observed in negative ionization mode:

891.1763
940.7420
996.1391
1058.4608
1129.0920
1209.8135
1302.9536
1411.6170
1540.0375
1694.1420
1882.4920
2117.9295
2420.6348

6 Perfluorokerosene [PFK, $CF_3(CF_2)_nCF_3$]

PFK can be analysed by electron impact and negative chemical ionization as a neat liquid. The following singly charged, monoisotopic ions are observed:

51.0046	342.9792	666.9601
68.9952	354.9792	680.9569
80.9952	366.9792	692.9569
92.9952	380.9760	704.9569
99.9936	392.9760	716.9569
111.9936	404.9760	730.9537
118.9920	416.9760	742.9537
130.9920	430.9728	754.9537
142.9920	442.9728	766.9537
149.9904	454.9728	780.9505
154.9920	466.9728	792.9505
168.9888	480.9696	804.9505
180.9888	492.9696	816.9505
192.9888	504.9696	830.9473
204.9888	516.9697	842.9473
218.9856	530.9664	854.9473
230.9856	542.9664	866.9473
242.9856	554.9664	880.9441
254.9856	566.9664	892.9441
268.9824	580.9633	904.9441
280.9824	592.9633	916.9441
292.9824	604.9633	930.9409
304.9824	616.9633	942.9409
318.9792	630.9601	954.9409
330.9792	642.9601	966.9409
	654.9601	

7 Polyethylene glycol (PEG)

PEG can be analysed by positive ionization thermospray, electrospray, and atmospheric pressure chemical ionization. A mixture of commercially available PEGs covering different mass ranges can be dissolved at a concentration of 0.1 mg mL^{-1} in 1:1 (v/v) acetonitrile–20 mM aqueous ammonium acetate. The ammonium acetate promotes ammoniated adducts. The following singly charged, monoisotopic, ions are observed for the species [(C$_2$H$_4$O)$_n$H$^+$], [(C$_2$H$_4$O)$_n$(H$_2$O)H$^+$], and [(C$_2$H$_4$O)$_n$(H$_2$O)NH$_4$$^+$], in positive ionization mode:

89.0603	432.2809	740.4644
133.0865	459.2805	767.4640
177.1127	476.3071	784.4906
195.1233	503.3068	811.4903
212.1498	520.3333	828.5168
239.1495	547.3330	855.5165
256.1760	564.3595	872.5430
283.1757	591.3592	899.5427
300.2022	608.3857	916.5692
327.2019	635.3854	943.5689
344.2284	652.4120	960.5955
371.2281	679.4116	987.5951
388.2547	696.4382	1004.6217
415.2543	723.4378	1031.6213

8 Sodium iodide–positive ions

Sodium iodide can be analysed by fast atom/ion bombardment and electrospray as an aqueous solution at a concentration of 1 mg mL^{-1}. Sometimes a trace amount of caesium iodide is added to supply the m/z 133 ions and bridge the rather large gap between m/z 23 and 173. The following singly charged, monoisotopic, [Na(NaI)$_n$$^+$] cluster ions are observed in positive ionization mode:

22.9898	1222.1437	2721.0861
(132.9054 – Cs$^+$)	1372.0379	2870.9803
172.8840	1521.9321	3020.8745
322.7782	1671.8264	3170.7688
472.6725	1821.7206	3320.6630
622.5667	1971.6149	3470.5572
772.4610	2121.5091	3620.4515
922.3552	2271.4033	3770.3457
1072.2494	2421.2976	3920.2400
	2571.1918	

9 Sodium iodide–negative ions

Sodium iodide can be analysed by fast atom/ion bombardment and electro-spray as an aqueous solution at a concentration of 1 mg mL^{-1}. The following singly charged, monoisotopic, $[I(NaI)_n{}^-]$ cluster ions are observed in negative ionization mode:

127.9045	2075.5296
276.7987	2225.4238
426.6929	2375.3180
576.5872	2525.2123
726.4814	2675.1065
876.3757	2825.0008
1026.2699	2974.8950
1176.1641	3124.7892
1326.0584	3274.6835
1475.9526	3424.5777
1625.8469	3574.4719
1775.7411	3742.3662
1925.6353	3874.2604

10 Tris(perfluoroalkyl)-*s*-triazines, (triazines)

Triazines of the type listed below can be analysed as supplied by electron impact, negative ionization chemical ionization, and fast atom/ion bombardment:

2,4,6-tris(perfluoroheptyl)-1,3,5-triazine $C_{24}F_{45}N_3$ MW 1185
2,4,6-tris(perfluorononyl)-1,3,5-triazine $C_{30}F_{57}N_3$ MW 1485

The following monoisotopic, singly charged ions are observed:

118.9920	470.9790	865.9581
125.9967	520.9758	889.9581
130.9920	544.9758	915.9550
137.9967	565.9774	927.9550
168.9888	570.9727	965.9516
175.9935	615.9742	1015.9485
180.9888	620.9694	1065.9451
218.9856	665.9710	1089.9453
230.9856	720.9630	1115.9421
268.9824	727.9677	1127.9421
280.9824	732.9630	1165.9390
318.9792	770.9599	1184.9373
375.9807	815.9613	1465.9300
420.9822	827.9613	1485.9283
	846.9597	

Subject Index